SPRINGER

LAB MANUALS

Springer Japan KK

Takehiko Koji (Ed.)

Molecular Histochemical Techniques

With 66 Figures, 6 in Color

Springer

TAKEHIKO KOJI, PH.D.
Associate Professor
Nagasaki University
Department of Histology and Cell Biology
1-12-4 Sakamoto
Nagasaki 852-8523
Japan

Library of Congress Cataloging-in-Publication Data
Molecular histochemical techniques / Takehiko Koji (ed). p. cm. – (Springer lab manual)
Includes bibliographical references and index.
ISBN 978-4-431-68520-3 ISBN 978-4-431-67915-8 (eBook)
DOI 10.1007/978-4-431-67915-8
1. In situ hybridization – Laboratory manuals. 2. Histochemistry – Laboratory manuals.
3. Nucleic acids – Analysis – Laboratory manuals. 4. MOlecular biology – Laboratory manuals.
I. Koji, Takehiko, 1955-II. Serie QH452.8 M65 2000 572.8-dc21

Originally published by Springer-Verlag Tokyo in 2000

Production: PRO EDIT GmbH, 69126 Heidelberg, Germany
Cover design: design & production GmbH, 69121 Heidelberg, Germany
Typesetting: Mitterweger & Partner GmbH, 68723 Plankstadt, Germany
SPIN: 10676803 05/3136 Re – 5 4 3 2 1 0

Preface

Historically, a variety of changes occurring in organisms were first recognized as morphological or histological ones and described with qualitative and visual terms, ie., with special anatomical and/or pathological terms. Therefore, even though there is a huge collection of morphological information on physiological as well as pathological states of cells, that information was highly descriptive and not always linked with molecular changes effectively. What is intended in this book is to provide newly developed histochemical methods with details, which should enable us to connect morphological information with molecular biological information.

The contents covered of this laboratory manual of molecular histochemical techniques show how to detect DNA strand breaks in situ (more familiarly, in individual cells or tissue sections), as an inevitable marker of apoptosis, how to detect specific sequences of DNA and RNA in situ, and how to detect specific DNA binding proteins like gene specific transcription regulatory factors in situ. All of them have been recently established and well recognized as essential techniques in actively growing fields of science beyond the histochemical world.

Of course, I invited the good standing, experienced scientists in practice to write the protocols, so that people who are interested in them have only to follow the protocols. Consequently, we are confident that the protocols in themselves or with a slight modification will soon be found to satisfy people in most cases.

It is our great pleasure to see what happens when using these molecular histochemical techniques which merge the classical morphological approach with the biochemical and molecular biological one. We hope this book can help those who want to know detailed methods on molecular histochemistry and also can shorten the time required for establishment of an experimental system of molecular histochemical studies in each laboratory.

Takehiko Koji, Ph. D.

Contents

Part I

Part I

Chapter 1

Introduction to the Detection of DNA Strand Breaks

TAKEHIKO KOJI

Introduction

DNA strand breaks, especially single-stranded breaks (SSB), are very commonly found through the life cycle of cells. In fact, SSB have been implicated to play important roles in the regulation of cell proliferation (Stewart et al.1990), cell differentiation (Appleby and Modak 1977; Farzaneh et al.1982; Dawson and Lough 1988) and cell death (Gold et al. 1993; Peitsch et al. 1993; Bortner et al. 1995; Hashimoto et al. 1995). Also, in the course of replicative DNA synthesis (Ogawa and Okazaki 1980), DNA repair (Ma et al. 1995) and genetic recombination in meiosis, SSB occur transiently. On the other hand, DNA double-stranded breaks (DSB) were often the results of DNA damages induced by irradiation of ionizing radial rays or by some antitumor drugs. Recently, however, DSB have been highlighted as a biochemical marker (Wyllie 1980) of apoptosis, the notion of which was introduced by Kerr et al. (1972). Currently, cell death is categorized into two groups; apoptosis and necrosis. Apoptosis, in contrast to necrosis, is morphologically characterized by the special features of chromatin condensation to the nuclear periphery and the fragmentation of the cell into apoptotic bodies. More importantly, apoptosis requires the de novo expression of specific genes and so implies that the death can be manipulated artificially.

To demonstrate the presence of SSB, the nucleoid sedimentation technique was previously undertaken as a biochemical method, where the nucleoids consisting of DNA in a supercoiled

---

Correspondence to Takehiko Koji, Nagasaki University, Dept. of Histology and Cell Biology, 1-12-4, Sakamoto, Nagasaki, 852-8523, Japan (*phone* +81-095-849-7025; *fax* +81-095-849-7028; *e-mail* tkoji@net.nagasaki-u.ac.jp)

conformation were analyzed on a sucrose-density gradient and the nucleoid sedimentation rate was determined as an indicator of the degree of SSB (Johnstone and Williams 1982). Although the presence of SSB in resting human peripheral blood lymphocytes was discovered by means of this method (Johnstone and Williams 1982; Greer and Kaplan 1984), the finding led to a big debate. In a short episode of whether the SSB are present in the nuclei of resting human peripheral blood lymphocytes or not, rightness of the nucleoid sedimentation technique was a major point in many conflicting reports. In fact, Jostes et al. (1989) and Boerrigter (1991) did not detect SSB in the DNA of the lymphocytes by means of alkaline elution method.

Besides, biochemical methods are generally not sufficient to correlate the occurrence of SSB with the changes in physiological states of each cell. Therefore, to understand better the biological significance of SSB and to get more direct evidence of a cause-and-effect relationship between SSB and cellular changes, a histochemical approach to localize the sites of SSB would be indispensable.

In the case of DSB detection, a ladder formation of DNA in agarose gel-electrophoresis is a most popular biochemical method and used especially often to show the presence of apoptotic cells. If about 4% of cell population are undergoing apoptosis (Collins et al. 1992), we can detect discrete ladder bands consisting of multiples of 180-200 base pairs. However, the method will not provide information on how many cells are apoptotic.

Histochemical methods provide useful information on the sites of DNA strand breaks at individual nuclear level and so we can get evidence of how many cells are haboring the strand breaks and where the positive cells are located. The combination with other histochemical methods to demonstrate the presence of various marker molecules may permit us to investigate the relation between occurrence of SSB or DSB of DNA and physiological changes at individual cell level.

In situ nick translation as a specific method to detect SSB

In order to analyze the occurrence of SSB directly at a level of individual cells, in situ nick translation (ISNT) could be a powerful tool (Koji 1996), which was originally developed to investigate the region of actively transcribed chromatin (Kerem et al. 1984) and performed with radioactive compounds. However, the radioactive ISNT is generally time-consuming and expensive, and it is hard to localize the sites of nicks precisely on the sub-nuclear basis. That is why, more recently, the application of non-radioactive ISNT has become frequent (Gold et al. 1994; Hashimoto et al. 1995). Furthermore, the needs for such a technique have recently increased rapidly in the particular field dealing with apoptotic cell death, which is characterized by a special occurrence of DNA strand breaks in the early stage of cell death commitment (Peitsch et al. 1993; Bortner et al. 1995). In addition, SSB are also known to be closely associated with necrosis in the cell death issues. Therefore, it now seems appropriate to give the experimental details in this methodology, as a part of molecular histochemical methods (Koji and Nakane 1996). In the following part, I would like to show the basic principle of nonradioactive ISNT and some examples in relation to the cell death subject.

"Nick translation" is a very common method of labeling cDNA to produce molecular biological probes, in which cDNA is nicked by DNase I digestion and at the nicked sites a new DNA strand is synthesized using the intact strand as a template. Basically, ISNT is based on the same principle to that of "nick translation". However, DNase I digestion is not needed.

The sites of SSB in nuclei can be first recognized by *E. coli* DNA polymerase I, which has three enzymatic activities; 5' to 3' DNA polymerase activity, 3' to 5' exonuclease activity and 5' to 3' exonuclease activity. As shown in Fig. 1, using the activities, the polymerase usually adds a new complementary deoxyribonucleotide to an intact free 3'-OH residue in the presence of dATP, dGTP, dCTP and TTP, depending upon the base sequence of the intact template strand. At the time of elongation, the newly synthesized DNA strand can be labeled with haptenic nucleotide analogues such as biotin-11-dUTP and digoxigenin-11-dUTP when the one of substrates, TTP is replaced with these analogues. Finally, the incorporated hapten molecules are reacted with horseradish peroxidase (HRP)-labeled anti-hapten antibody and visualized enzyme-immunohistochemically. In "nick translation", the polymerase reaction is usually conducted around 15° C to ensure the high fidelity of the synthesized base sequences to the template strand. In ISNT, however, such a high fidelity in base complementarity between a newly synthesized

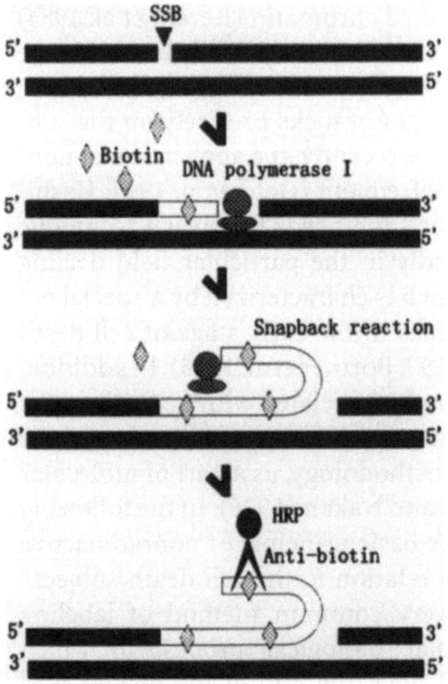

Fig. 1. Principle of nonradioactive in situ nick translation (ISNT) (Koji 1996). For details, see the text.

strand and the template strand is not required. Therefore, we could accomplish a remarkable enhancement of detection sensitivity of ISNT by conducting the reaction at 37° C, which allows to generate the "snapback" synthesis of DNA. In addition, the pretreatment of sections with proteinase K is essentially important, particulary to discriminate SSB between apoptosis and necrosis by this method (Koji 1996; Hashimoto et al. 1995, 1997).

For staining examples of ISNT, I will show the occurrence of SSB during DNA repair process after UV irradiation of human peripheral blood lymphocytes (PBL). As shown in Fig.2, no detectable level of SSB was observed in the normal resting PBL, irrespective of the protease treatment. Two hrs after UV irradiation, the signal for SSB appeared in the nuclei of PBL and the staining was not dependent upon proteinase K digestion, while the protease treatment resulted in an increase in the intensity of the staining. One day after UV irradiation, most of the staining for SSB disappeared. As stated above, there were many conflict-

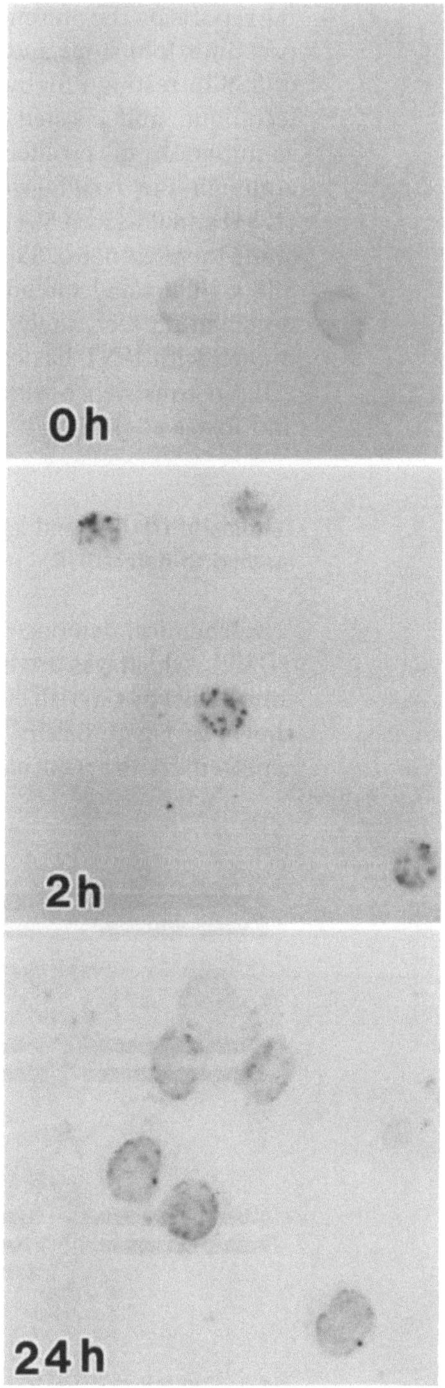

Fig. 2. SSB detected by ISNT in human peripheral blood lymphocytes after UV irradiation (Koji 1996). Human blood was drawn from healthy male volunteers and separated in Ficoll-Paque density-gradient centrifugation. The lymphocyte fraction was collected and washed with PBS. Finally, the PBL were suspended in 0.25 M sucrose. An aliquot of cell suspension was irradiated with UV at a dose of 10 J/m^2, and then the cells were washed with MEM. The final pellet was resuspended in RPMI 1640 medium supplemented with 10% fetal bovine serum and allowed to repair UV-induced DNA damage at 37° C under 5% CO_2 in air. At various time-points, an aliquot of cell suspension was centrifuged onto gelatin-coated glass slides, as detailed previously (Koji 1996). The sites of SSB were detected by ISNT as black spots. For more details, see Chapter 2.

ing reports on the presence of SSB in resting human PBL. For the first time, Johnstone and Williams (1982) reported the presence of SSB in resting PBL by means of the nucleoid sedimentation technique, and insisted that the SSB must be ligated prior to re-initiation of proliferation in response to lectins. Later, although the results were confirmed by Greer and Kaplan (1984) and Prasad et al. (1987), the other groups failed to demonstrate the presence of SSB by different methods, arguing that the SSB are illusional and an artifactual product of the nucleoid sedimentation technique. As shown here, the histochemical approach with ISNT has clearly presented the absence of SSB in PBL, in consistence with the results of Boerrigter et al. (1991) and Jostes et al. (1989).

Terminal dUTP nick end labeling (TUNEL) as a preferential method to detect DSB

Histochemical demonstration of DSB of DNA is carried out by TUNEL, which was originally reported as a tool to detect apoptotic nuclei by Gavrieli et al. (1992). The principle of TUNEL was shown in Fig. 3. Briefly, terminal free 3'-OH of DNA is added repeatedly with a nucleotide by terminal deoxynucleotidyl trans-

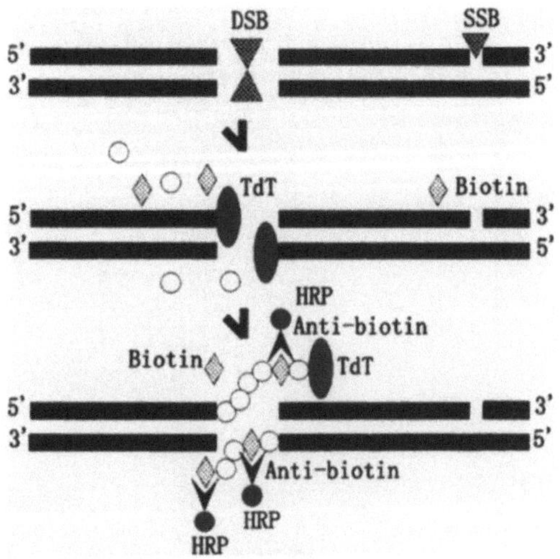

Fig. 3. Principle of TUNEL. For details, see the text.

ferase (TdT). When a tissue section is reacted with TdT in the presence of biotin-16-dUTP or digoxigenin-11-dUTP, the sites of DNA strand breaks can be labeled with these haptenic nucleotide analogues. Finally, the haptens are detected immunohistochemically with enzyme-labeled antibodies such as HRP-labeled anti-biotin.

In tissue sections, differing from the case in solution, the sites of DSB are preferentially reacted with the enzyme compared to that of SSB. Thus, TUNEL is now the most frequently used method to detect apoptotic cells which have abundant DSB of DNA. Here, however, I would like to raise a warning that TUNEL does not always detect apoptotic cells. As Bortner et al. (1995) reviewed, it is not prerequisite for apoptotic cells to show a significant level of DNA double-stranded breaks at an internucleosomal site. It is also significant that TdT in itself is not specific to the 3'-OH terminus of DSB, rather it can even react efficiently with

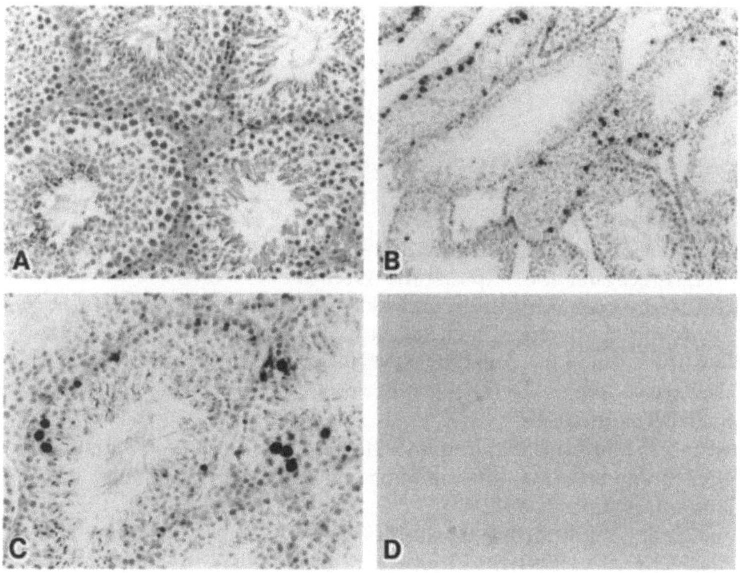

Fig. 4. TUNEL staining of normal adult mouse testis. Paraffin-embedded sections of adult mouse testis fixed with 4% paraformaldehyde in PBS (pH 7.4) were mounted to silane-coated glass slides and used for TUNEL staining after appropriate proteinase K digestion (See Chapter 3 for more details). A: Hematoxylin-eosin staining. B: Low-power view of TUNEL positive cells. C: High-power view of TUNEL possitive cells. D: Negative control without TdT.

that of single-stranded DNA. That is why we have been using TdT to label synthetic oligodeoxynucleotides as detailed in the protocol of in situ hybridization (Koji and Brenner 1993; Chapter 12 and 13). In fact, a specific nuclear staining was obtained by TUNEL in the frozen sections of CCl_4 treated liver, where almost all of degenerating hepatocytes were observed with typical necrotic features by electron microscopy (Hashimoto et al. 1995). In addition, Gold et al. (1994) reported that early stages of necrosis were preferentially detected by ISNT, whereas TUNEL was slightly more sensitive for the detection of apoptosis. In our hands, it has turned out that TUNEL is specific enough to detect apoptotic germ cells in mouse testis (Wang et al. 1998) and apoptotic renal tubular cells in ischemia-reperfusion mouse kidney (Nogae et al. 1998), as verified by combining use of electron microscopy. Consequently, we should realize that the usefulness of TUNEL in identifying apoptotic cells seems highly dependent upon the tissue-type or cell-type to be analyzed. An example of TUNEL staining of apoptotic spermatogenic cells in adult mouse testis was shown in Fig. 4 (Koji and Wang 1998).

References

Appleby DW, Modak SP (1977) DNA degradation in terminally differentiating lens fiber cells from chick embryos. Proc Natl Acad Sci USA 74:5579-5583

Boerrigter METI (1991) DNA strand break metabolism in human lymphocytes: A reevaluation. Exp Cell Res 196:1-5

Bortner CD, Oldenburg NBE, Cidlowski JA (1995) The role of DNA fragmentation in apoptosis. Trends Cell Biol 5:21-26

Collins RJ, Harmon BV, Gobe GC, Kerr JFR (1992) Internucleosomal DNA cleavage should not be the sole criterion for identifying apoptosis. Int J Radiat Res 61:451-453

Dawson BA, Lough J (1988) Immunocytochemical localization of transient DNA strand breaks in differentiating myotubes using in situ nick-translation. Dev Biol 127:362-367

Farzaneh F, Zalin R, Brill D, Shall S (1982) DNA strand breaks and ADP-ribosyl transferase activity in eukaryotic differentiation. Nature 300:362-366

Gavrieli Y, Sherman Y, Ben-Sasson SA (1992) Identification of programmed cell death in situ via specific labeling of nuclear DNA fragmentation. J Cell Biol 119:493-501

Gold R, Schmied M, Breitschopf H, Hartung HP, Toyka KV, Lassmann H (1994) Differentiation between cellular apoptosis and necrosis by the combined use of in situ tailing and nick translation techniques. Lab Invest 71:219-225

Greer WL, Kaplan JG (1984) Regulation of repair of naturally occurring DNA strand breaks in lymphocytes. Biochem Biophys Res Commun 122:366-372

Hashimoto S, Koji T, Niu J, Kanematsu T, Nakane PK (1995) Differential staining of DNA strand breaks in dying cells by non-radioactive in situ nick translation. Arch Histol Cytol 58: 161-170

Hashimoto S, Koji T, Kohara N, Kanematsu T, Nakane PK (1997) Frequency of apoptosis relates inversely to invasiveness and metastatic activity in human colorectal cancer. Virchows Arch 431:241-248

Johnstone AP, Williams GT (1982) Role of DNA breaks and ADP-ribosyl transferase activity in eukaryotic differentiation demonstrated in human lymphocytes. Nature 300:368-370

Jostes R, Reese JA, Cleaver JE, Molero M, Morgan WF (1989) Quiescent human lymphocytes do not contain DNA strand breaks detectable by alkaline elution. Exp Cell Res 182:513-520

Kerem B, Goitein R, Diamond G, Cedar H, Marcus M (1984) Mapping of DNase-I sensitive regions on mitotic chromosomes. Cell 38:493-499

Kerr JFR, Wyllie AH, Currie AR (1972) Apoptosis: A basic biological phenomenon with wide-ranging implications in tissue kinetics. Brit J Cancer 26:239-257

Koji T (1996) Nonradioactive in situ nick translation: A useful molecular histochemical tool to detect single-stranded DNA breaks. Acta Histochem Cytochem 29:71-79

Koji T, Brenner RM (1993) Localization of estrogen receptor messenger ribonucleic acid in rhesus monkey uterus by nonradioactive in situ hybridization with digoxigenin-labeled oligodeoxynucleotides. Endocrinology 132:382-392

Koji T, Nakane PK (1996) Recent advances in molecular histochemical techniques: In situ hybridization and southwestern histochemistry. J Electron Microsc 45:119-127

Koji T, Wang R-A (1998) Male germ cell death in mouse testes. In: Yamada T and Hashimoto Y (ed) Apoptosis: Mechanism and Function. Business Ctr Acad Soc Japan, Tokyo, p.85-96

Ma L (1995) Mammalian nucleotide repair. Biochim Biophys Acta 1242: 137-164

Nogae S, Miyazaki M, Kobayashi N, Saito T, Abe K, Saito H, Nakane PK, Nakanishi Y, Koji T (1988) Induction of apoptosis in ischemia-reperfusion model of mouse kidney: Possible involvement of Fas. J Am Soc Nephrol,9:620-631

Ogawa T, Okazaki T (1980) Discontinuous DNA replication. Annu Rev Biochem 49: 421-457

Peitsch MC, Muller C, Tschopp J (1993) DNA fragmentation during apoptosis is caused by frequent single-strand cuts. Nucl Acids Res 21:4206-4209

Prasad KV, Greer WL, Severini A, Kaplan JG (1987) Increase in intercellular Na^+: Transmembrane signal for rejoining of DNA strand breaks in proliferating lymphocytes. Cancer Res 47:5397-5400

Stewart AF, Herrara RE, Nordheim A (1990) Rapid induction of c-fos transcription reveals quantitative linkage of RNA polymerase II and DNA topoisomerase I enzyme activities. Cell 60:141-149

Wang R-A, Nakane PK, Koji T (1998) Autonomous cell death of mouse germ cells during fetal and postnatal period. Biol Reprod, 58:1250-1256

Wyllie AH (1980) Glucocorticoid-induced thymocyte apoptosis is associated with endogenous endonuclease activation. Nature 284:555-556

Chapter 2

In Situ Nick Translation

SATOSHI HASHIMOTO AND TAKASHI KANEMATSU

Introduction

Nick (DNA single strand break) have been implicated to play important roles in the regulation of cell proliferation (Ogawa 1980), cell differentiation (Appleby 1977) and cell death (Peitsch 1993). Nick translation detects nicks in DNA by adding nucleotides to 3'-hydroxyl ends in the presence of a template, extending the strand in the 5' to 3' direction. (Fig. 1) In situ nick translation (ISNT)(Dawson 1980) is application of nick translation on frozen sections or formalin fixed paraffin embedded tissues.

Outline

1. Preparation of samples
2. Nick translation reaction
3. Blocking and immunohistochemistry
4. Visualization

Materials

- 3-aminopropyltriethoxysilane (Sigma, USA) coated slides

Correspondence to Satoshi Hashimoto, Nagasaki University School of Medicine, Second Department of Surgery, 1-7-1, Sakamoto, Nagasaki, 852, Japan (*phone* +81-95-849-7316; *fax* +81-95-849-7319; *e-mail* hashis@mb.-infoweb.ne.jp); Takashi Kanematsu, Nagasaki University School of Medicine, Second Department of Surgery, 1-7-1, Sakamoto, Nagasaki, 852, Japan

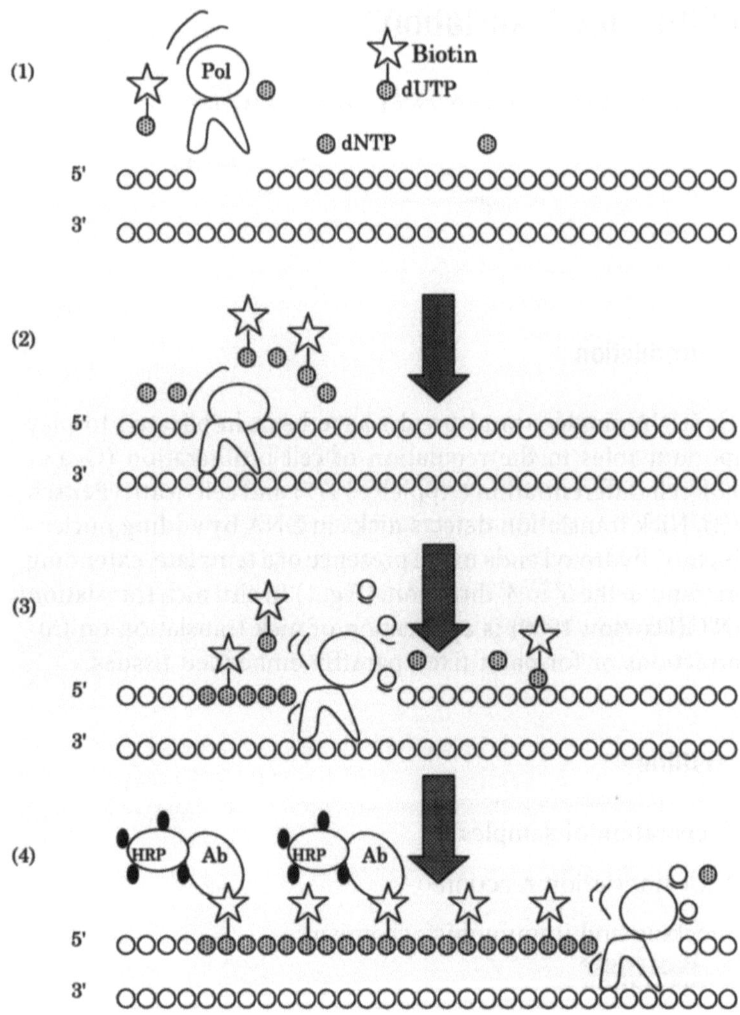

Fig. 1. Principle of non-radioactive ISNT. (1) First nick contacts with NT buffer and *E. coli* DNA polymerase I. (2) *E. coli* DNA polymerase I first recognize "nick" and connect with it. (3) The polymerase usually adds a new complementary deoxynucleotide. Newly synthesized DNA strand can be labeled with haptenic nucleotide analogues such as biotin-11-dUTP. (4) Biotin are reacted with horseradish peroxydase (HRP)-labeled anti-biotin antibody and visualized enzyme-immunohistochemically. *Pol*: *E. coli* DNA polymerase I *Ab*: Anti-biotin antibody Inc., USA), *HRP*: Horseradish peroxydase

Samples

- Fresh frozen samples: Tissues are embedded in OCT compound (Miles, USA), cut(5-6μm), placed onto silane coated glass slides and air dried. Then sections are fixed with ethanol/acetic acid (3:1, v/v) for 20 mins, and washed 3 times with PBS for 5 mins each time.
- Formalin-fixed and paraffin embedded samples: Tissues are cut (4 μm), placed onto silane-coated glass slides and deparaffined.
- Proteinase K (PK) (Sigma, USA)
- Biotin-11-dUTP (ENZO Diagnostics, Inc., USA)
- *E.coli* DNA polymerase I (Takara Co., Japan)
- Horseradish peroxydase (HRP) labeled anti-biotin antibody (Vector Lab. Inc., USA)

NT buffer:
- 50 mM Tris/HCl (pH 7.5)
- 10 mM $MgCl_2$
- 0.1 mM dithiothreitol
- 50 μg/ml BSA
- 20 μM dATP, 20 μM dGTP and 20 μM dCTP
- 20 μM TTP(negative control) or 20 μM biotin-11-dUTP

Chromogenic solution (Adams 1981):
- 0.1 M sodium phosphate buffer (pH 7.5)
- 0.5 mg/ml 3-3'-diaminobenzidine/4HCl
- 0.01% hydrogen peroxide
- 0.025% $CoCl_2$
- 0.02% $NiSO_4(NH_4)_2SO_4$

Procedure

1. Immerse in PBS.
2. If necessary, PK digestion for 15 mins at 37°C and washing in PBS.
   (Conc. of PK is recommended at 1 μg/ml for fresh frozen sample and 10 μg/ml for paraffin embedded samples.)
3. Keep in 50 mM Tris/HCl.
4. Nick translation at 37 °C for 3 hrs in a moist chamber Reaction with *E.coli* DNA polymerase I(200 U/ml) in NT buffer.

5. Wash 3 times with 50 mM Tris/HCl for 5 mins, each time and rinse once with PBS.
6. Blocking with 500 μg/ml normal goat IgG and 5% BSA/PBS at RT for 1 hr.
7. Reaction with HRP-goat anti Biotin for 3 hrs.
8. Wash with 0.075% Brij./PBS for 15 mins, 4 times.
9. Visualize the sites of HRP by immersing in chromogenic solution for 5-6 mins.
10. Dehydration with serial upgraded ethanol solutions, clear with xylene and mount without counterstaining.

Results

Nicks are visualized as intense signal in nuclei (Fig. 2 a,b,c). Signal is varied from fixation or proteinase treatment of samples (Koji 1996)(Hashimoto 1995). As a negative control, some sec-

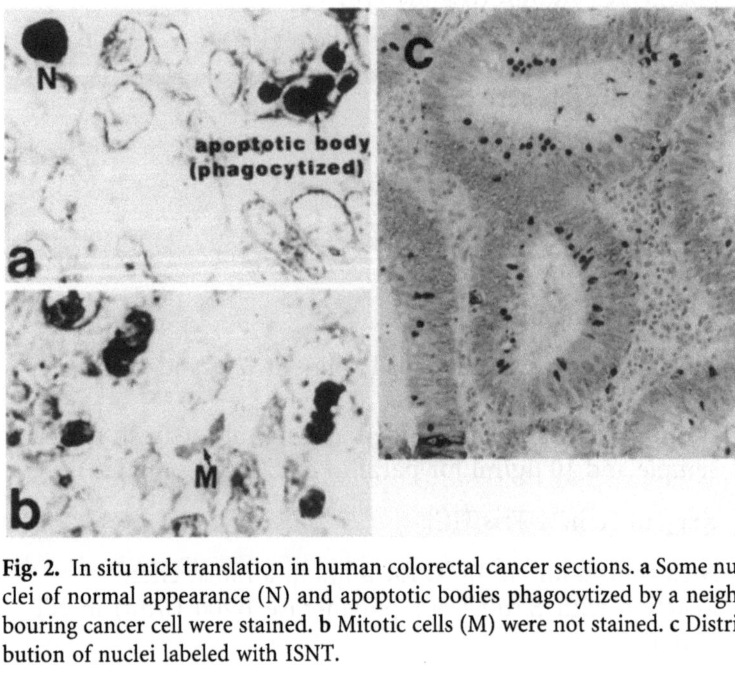

Fig. 2. In situ nick translation in human colorectal cancer sections. **a** Some nuclei of normal appearance (N) and apoptotic bodies phagocytized by a neighbouring cancer cell were stained. **b** Mitotic cells (M) were not stained. **c** Distribution of nuclei labeled with ISNT.

tions in the absence of DNA polymerase and/or biotinylated dUTP which inform specificity of the reaction should be added and their signals are essentially blank.

Troubleshooting

Heavy proteinase treatment results in a high background staining and a destruction of the samples especially in their edge.

First, the optimal conc. of PK is determined according to the fixation or organ of the tissues respectively. And PK digestion should be performed at 37 °C not at RT.

Comments

We have applied this technique to characterize human colorectal cancers (Hashimoto 1997). In paraffin sections, nuclei labeled with ISNT were different from Proliferating cell nuclear antigen positive nuclei. DNA obtained from cancer tissues that contained many ISNT positive nuclei electrophoresed as a ladder pattern. Nuclei labeled with ISNT coincide with the morphologic apoptosis and hallmark of apoptosis.

And in frozen sections, ISNT is a convenient molecular histochemical tool not only to detect cell death but also to discriminate both types of cell death, apoptosis and necrosis (Hashimoto 1995).

References

Adams JC (1981) Heavy metal intensification of DAB-based HRP reaction product. J Histochem Cytochem 29: 775

Appleby DW, Modak SP (1977) DNA degradation in terminally differentiating lens fiber from chick embroys. Proc Natl Acad Sci USA 74: 5579-5583

Dawson BA, Lough J (1988) Immunocytochemical localization of transient DNA strand breaks in differentiating myotubes using in situ nick-translation. Dev Biol 127:362-367

Hashimoto S, Koji T, Niu J, Kanematsu T, Nakane PK (1995) Differential staining of DNA strand breaks in dying cells by non-radioactive in situ nick translation. Arch Histol Cytol 58:161-170

Hashimoto S, Koji T, Kohara N, Kanematsu T, Nakane PK (1997) Frequency of apoptosis relates inversely to invasiveness and metastatic activity in human colorectal cancer. Virchows Arch 431:241-248

Koji T (1996) Non-radioactive in situ nick translation : A useful molecular histochemical tool to detect single-stranded DNA breaks. Acta Histochem Cytochem 29:71-79

Ogawa T, Okazaki T (1980) Discontinuous DNA replication. Annu Rev Biochem 49: 421-457

Peitsch MC, Muller C, Tschopp J (1993) DNA fragmentation during apoptosis is caused by frequent single-strand cuts. Nucl Acids Res 21:4206-4209

Chapter 3

In Situ Nick End-Labeling: Light Microscopical

SHIGEYOSHI OBA AND TAKEHIKO KOJI

Introduction

It has been postulated that in apoptotic cell death, the first biochemical event is single-strand DNA breaks followed by double-strand DNA cleavage in the linker regions between nucleosomes (Peitsh 1993). TUNEL (Terminal transferase-mediated dUTP-biotin nick end labeling) is a convenient molecular histochemical tool to detect these single-DNA breaks. Briefly described, this tool is based on repeated addition of biotin-dUTP at the 3-hydroxy ends of DNA of tissue by terminal deoxynucleotidyl transferase and detection of these additional biotin-dUTP by HRP labeled anti-biotin antibody. You can use digoxigenin-dUTP or fluorescent conjugated dUTP instead of biotin-dUTP.

Outline

The entire protocol of TUNEL is outlined in Fig. 1 and Fig. 2.

Materials

- Proteinase K (Boehringer Mannheim Biochemicals, Germany or Sigma Chemical Co., U.S.A)

Correspondence to Shigeyoshi Oba, The University of Tokyo, The 2nd Department of Internal Medicine, Faculty of Medicine, 3-7-1 Hongo, Bunkyo-ku, Tokyo, Japan (*phone* +81-3-3815-5411 (ext. 3056); *fax* +81-3-3814-0021; *e-mail* oba-2im@h.u-tokyo.ac.jp); Takehiko Koji, Nagasaki University, Dept. of Histology and Cell Biology, 1-12-4, Sakamoto, Nagasaki, 852-8523, Japan

Deparaffinization and rehydrate

↓

PBS 5mins

↓

Proteinase K 15mins at 37°C

↓

PBS 5minsX3

↓

Preincubate with TdT buffer RT 30mins

↓

TdT reaction 37°C 2hrs

↓

Tris HCl rinse 5minsX3

↓

0.3% H_2O_2 RT 15mins

↓

Rinse Brij/PBS 5minsX3

↓

Normal goat IgG RT 1hr

↓

HRP-conjugated goat anti-biotin RT 2hrs

↓

Rinse Brij/PBS 10minsX4

↓

DAB-H_2O_2 with $CoCl_2$ and $NiSO_4(NH_4)_2SO_4$ RT 5mins

↓

Rinse distilled water > 5mins

↓

Dehydration and mounting

Fig. 1. Paraffin block with biotin-16-dUTP

Air dry cryostat sections(5-8μm)

↓

PBS 5mins

↓

Fix with 4% paraformaldehyde 20mins

↓

PBS 5minsX3

↓

Proteinase K 15mins at 37°C

↓

PBS 5minsX3

↓

Preincubate with TdT buffer RT 30mins

↓

TdT reaction 37°C 2hrs

↓

Tris HCl rinse 5minsX3

↓

0.3% H_2O_2 RT 5mins

↓

Rinse Brij/PBS 5minsX3

↓

Normal sheep IgG RT 1hr

↓

HRP-conjugated sheep anti-digoxigenin RT 2hrs

↓

Rinse Brij/PBS 10minsX4

↓

DAB-H_2O_2 with $CoCl_2$ and $NiSO_4(NH_4)_2SO_4$ RT 5mins

↓

Rinse distilled water > 5mins

↓

Dehydration and mounting

Fig. 2. Frozen sections with digoxigenin-11-dUTP

- Terminal deoxynucleotidyl transferase (TdT), TdT reaction buffer and $CoCL_2$ (Boehringer Mannheim Biochemicals, Germany)
- Biotin-16-dUTP (Boehringer Mannheim Biochemicals, Germany)
- digoxigenin-11-dUTP (Boehringer Mannheim Biochemicals, Germany)
- HRP labeled goat anti-biotin antibody (Vector Lab. Inc., U.S.A.)
- HRP labeled sheep anti-digoxigenin antibody (Boehringer Mannheim Biochemicals, Germany)
- Normal goat IgG (Dako, Copenhagen, Denmark)
- Normal sheep IgG (Dako, Copenhagen, Denmark)
- Bovine serum albumin (BSA)(Sigma chemical Co., U.S.A.)
- Dithiothreitol (DTT) (Sigma chemical Co., U.S.A.)
- deoxyadenosine triphosphate (dATP) (Sigma chemical Co., U.S.A.)
- Brij 35 (Sigma chemical Co., U.S.A.)
- 0.075% (v/v) Brij35/PBS; 5ml Brij35/2000ml PBS

1x TdT buffer (total 1000 μl)

distilled water	800 μl
5x reaction buffer (Boehringer)	200 μl

Terminal transferase kit (Boehringer Mannheim Biochemicals, Germany; No. 220582) contains 5X reaction buffer, $CoCl_2$ and TdT

TdT reaction solution (total 1000μl)-biotin

distilled water	701μl
5X reaction buffer (Boehringer Industries)	200μl
DTT 10 mM	10μl (0.1mM; final concentration)
$CoCl_2$ 25mM (Boehringer Industries)	60μl (1.5mM)
dATP 1mM	20μl (20μM)
biotin-16-dUTP (Boehringer Industries, 1mM)	5μl (5μM)
TdT(Boehringer Industries, 25U/ml)	4μl (0.1U/μl)

Terminal transferase kit (Boehringer Mannheim Biochemicals, Germany; No. 220582) contains 5X reaction buffer, $CoCl_2$ and TdT You may use lower volume of biotin-16-dUTP when there is excessive background staining in your sample (Fig. 4). You had better stock DTT (10mM) and dATP(1mM)/ distilled water below -20°C

TdT reaction solution (total 1000μl)-digoxigenin

distilled water	701μl
5X reaction buffer (Boehringer Industries)	200μl
DTT 10mM	10μl (0.1mM)
$CoCl_2$ 25mM (Boehringer Industries)	60μl (1.5mM)
dATP 1mM	20μl (20μM)
digoxigenin-11-dUTP (Boehringer Industries,1mM)	5μl (5μM)
TdT (Boehringer Industries, 25U/ml)	4μl (0.1U/μl)

Terminal transferase kit (Boehringer Mannheim Biochemicals, Germany; No. 220582) contains 5X reaction buffer, $CoCl_2$ and TdT

3,3'-diaminobenzidine(DAB) reaction solution

distilled water	90 ml
DAB	50 mg
1 M phosphate buffer, pH 1.2	10 ml
$CoCl_2$ (1%)	2.5 ml
$NiSO_4$ (1%)	2.0 ml
30% hydrogen peroxide (H_2O_2)	33μl

In our protocol, the sites of horseradish peroxides(HRP) were visualized with hydrogen peroxide and 3,3'-diamnobenzidine (DAB) in the presence of nickel and cobalt irons, which increase the sensitivity more than tenfold as compared to that of hydrogen peroxide and DAB alone (Adams 1981).

Procedure

Paraffin sections with biotin-dUTP

1. Samples are fixed in 4% paraformaldehyde in phosphate-buffered saline (pBS), pH 1.4 at RT about 4-10hrs.
2. Samples are rinsed with PBS 3times for 5mins each, dehydrated and embedded in paraffin block by routine method.
3. Cut sections to a thickness of 5-6μm and place them on silane-coated glass slides according routine procedure.
4. Deparaffinize the sections with toluene 3times (5 mins in each cup), and rehydrate through graded ethanol.
5. Wash with distilled water for 5 mins, 2 times, then immerse in PBS for 5 mins.
6. Proteinase K treatment; Incubate the sections in the proteinase K solution(0.5 -10μg/ml in PBS) at 37°C for 15mins.

Note: You must change the concentration of proteinase K according to sample and fixation condition. You had better try several concentration of proteinaseK for first time(for example 0.1, 0.5, 1, 5, 10μg/ml) and decide on the most appropriate concentration for your sample.
(Put the sections into the cup in 37°C water bath, which contains 100μl-2ml PK(500μg/ml in PBS) and 100ml PBS.)

Note: You had better stock higher concentration of proteinase K(100 or 500μg/ml in PBS) below -20°C.

7. Wash with PBS 3 times, 5 mins each
8. Rinse once with distilled water.
9. Preincubate with 1×TdT buffer, which contains 0.2M potassium cacodylate, 25 mM Tris-HCl(pH 6.6) and 0.025%BSA for 30mins at room temperature(RT). (Tap off excess water, carefully wipe around specimen and apply TdT buffer to cover specimen in moist chamber)
10. Incubate in 1x TdT buffer at 37°C for 2hrs in moist chamber.

(Tap off excess buffer, wipe around specimen and apply enough TdT reaction solution -biotin to cover specimen in moist chamber at 37°C)

TdT reaction solution - biotin

11. Rinse the samples 3 times with 50mM PH 7.5 Tris-HCl for 5 mins each.

12. Immerse the samples in 0.3% H_2O_2 in methanol for 15 mins at RT to inactivate the endogeneous peroxidase. (Immerse the samples into the cup which contains 1ml 30%H_2O_2 and 99ml methanol)

13. Wash three times with 0.075% (v/v) Brij35/PBS for 5 mins each.

Note: In this protocol, H_2O_2 and antibody are washed with 0.075% Brij35/PBS which increase the efficiency of washing as compared to only PBS.

14. After a brief washing with PBS, incubate with normal goat IgG (0.5 mg/ml in 5% BSA/PBS) at RT for 1hr to reduce the nonspecific binding of antibody to the samples.(Tap off excess buffer, wipe around specimen and apply normal goat IgG to cover specimen in moist chamber at RT)

15. Incubate with HRP-conjugated goat anti-biotin (1/100 in 1%BSA/PBS) at RT for 2hrs. (Tap off excess normal gout IgG solution, wipe around specimen and apply enough HRP-conjugated goat anti-biotin to cover specimen in moist chamber at RT).

16. Wash with 0.075% (v/v) Brij35/PBS 4 times, 10 mins each, then immerse in PBS.

17. Visualize with DAB-H_2O_2 and enhance with $CoCl_2$ and $NiSO_4(NH_4)_2SO_4$ at RT for 5mins. (Immerse the samples into the cup containing (DAB) reaction solution at RT for 5mins).

DAB reaction solution

18. Wash with running water for short time and finally distilled water for 5 mins.

19. Dehydration and mounting.

Frozen section with digoxigenin-dUTP

1. Cryostat sections (5-8 μm) are cut from snap-frozen tissue blocks. and air dried for 2-4hours.
2. Wash dried sections with PBS and fix in the 4% paraformaldehyde in PBS for 20 mins at RT.
3. Wash the samples with PBS for 3 times, 5 min each
4. Proteinase K treatment; Incubation the sections in the Proteinase K solution(0.5-1.0μg/ml in PBS) at 37°C for 15min.

Note: You had better use lower concentration of proteinaseK for frozen sections than paraffin sections.

5. Wash with PBS 3 times, 5 mins each
6. Rinse once with distilled water.
7. Preincubate with 1×TdT buffer, for 30mins at room temperature(RT).

TdT reaction solution - digoxigenin

8. Incubate in TdT reaction solution -digoxigenin 37°C for 2hrs in moist chamber.
9. Rinse the samples 3 times with 50mM PH 7.5 Tris-HCl for 5 mins each.
10. Immerse the samples in 0.3% H_2O_2 in methanol for 15 mins at RT to inactivate the endogeneous peroxidase.
11. Wash three times with Brij/PBS for 5 mins each.
12. After a brief washing with PBS, incubate with normal sheep IgG(0.5 mg/ml in 5% BSA/PBS) at RT for 1hr to reduce the nonspecific binding of antibody to the samples.
13. Incubate with HRP-conjugated sheep anti-digoxigenin (1/100 in 1%BSA/PBS) at RT for 2hrs.
14. Wash with 0.075% Brij/PBS 4 times, 10 mins each, then immerse in PBS.
15. Visualize with DAB-H_2O_2 and enhance with $CoCl_2$ and $NiSO_4(NH_4)_2SO_4$ at RT for 5mins.
16. Wash with running water for short time and finally distilled water for 5 mins.
17. Dehydration and mounting.

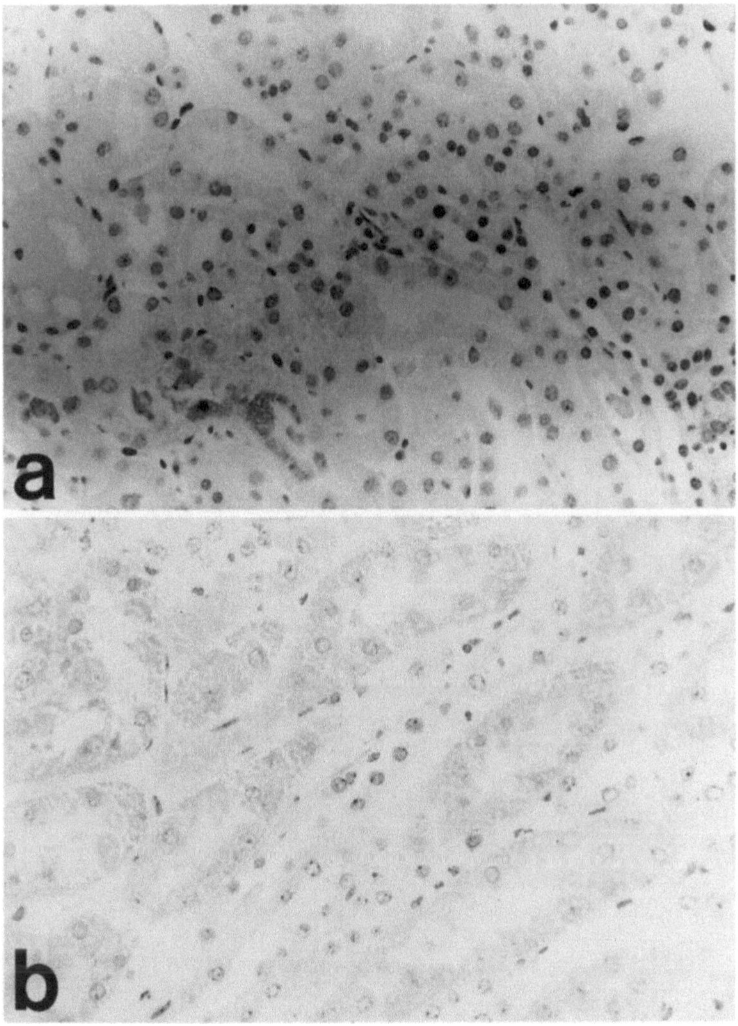

Fig. 3. TUNEL of ischemia-reperfusion kidneys(concentration of proteinase K). Samples treated with 10μg/ml proteinase K (**a**) are reacted with terminal deoxynucleotidyl tranferase reasonably, although samples treated with 1.0 μg/ml proteinae K (**b**) can not be reacted well.

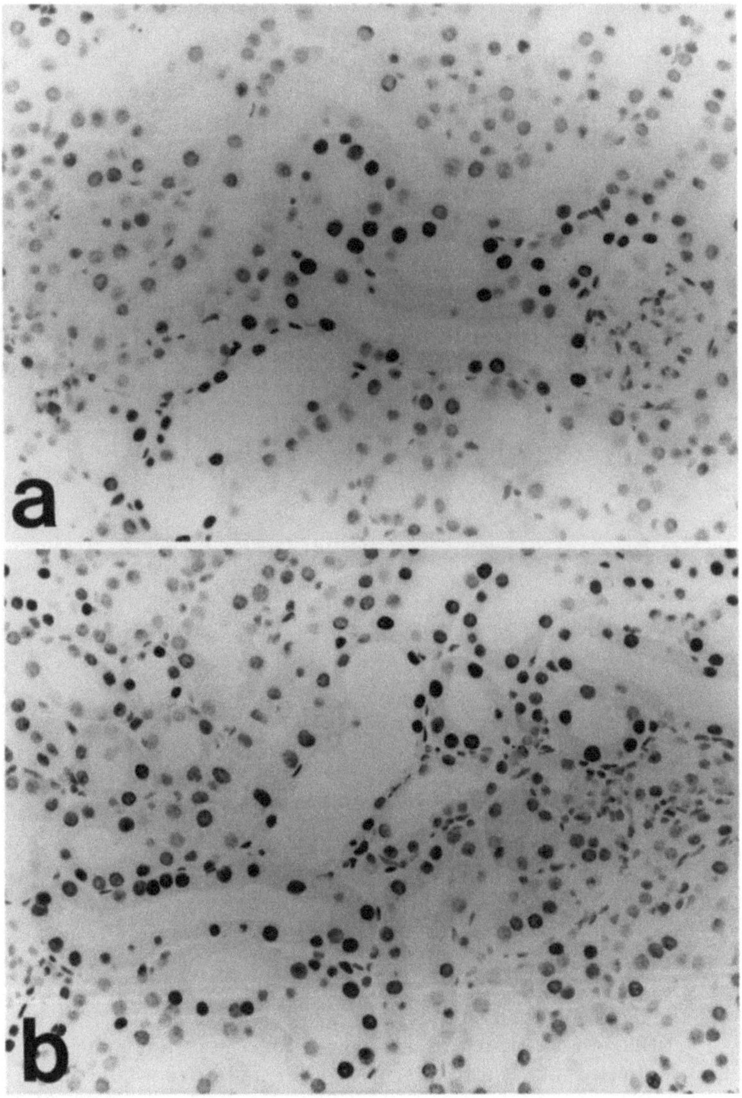

Fig. 4. TUNEL of ischemia-reperfusion kidneys(concentration of biotin-16 dUTP). The samples treated with TdT reaction solution containing 0.25μM biotin-16-dUTP stain moderately (**a**), although there is excessive background staining in the samples treated with 1 (**b**), 2.5 (**c**), 5μM (**d**) biotin-16-dUTP.

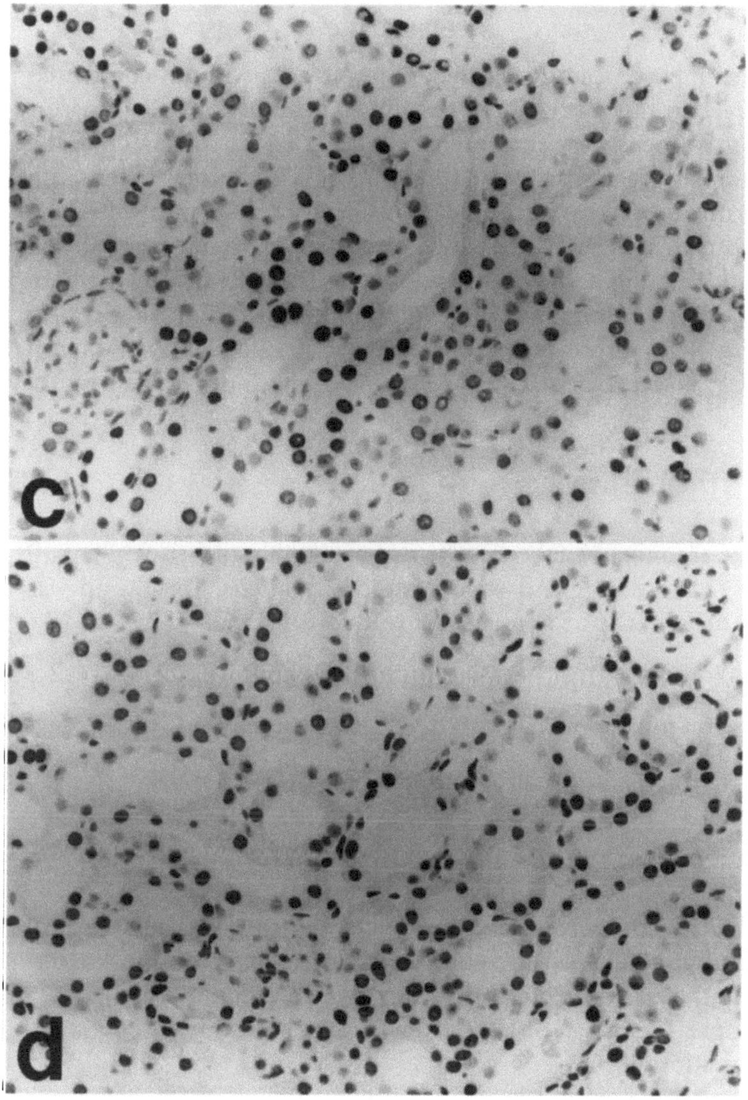

Fig. 4. c–d

Results

This method can detect the apoptotic cells without condensed nuclei in situ. For examples, apoptotic cells of ischemia-reperfusion kidneys were shown Fig. 3 and Fig. 4. Fig. 3 shows that the samples treated with 10μg/ml Proteinase K are reacted with terminal deoxynucleotidyl transferase reasonably, although those treated with 1.0μg/ml Proteinase K are not reacted well with TdT. Fig. 4 shows that the samples treated with TdT reaction solution containing 0.25μM biotin-16-dUTP stain moderately (Fig. 4 **a**), although there is excessive background staining in the samples treated with 1, 2.5, 5μM biotin-16-dUTP(Fig 4 **b,c,d**).

Troubleshooting

- Care must be taken not to dry up samples, when you tap off excess buffer and wipe around specimens. Drying up was main cause of non specific reaction.
- You must treat negative controls without TdT.
- You must treat TdT carefully because it may easily lose activity below or up -20°C.

Comments

This method can detect apoptosis cell at early phase but it can not discriminate between apoptosis and necrosis. You must ascertain the existence of apoptosis by using another method, for example, formation of DNA laddering, electron microscoptic observation, FACS.

References

Adams JC.; J. Histochem Cytochem 29: 775, 1981

Gavieli, Y., Sherman, Y., Ben-Sasson, S.A.; J. Cell Biol., 119:493-501,1992

Peitsh MC, Muller C, Tschopp J (1993) DNA fragmentation during apoptosis is caused by frequent single-strand cuts. Nucl. Acid Res. 21:4206-4209

Chapter 4

In Situ Nick End-Labeling: Electron Microscopical

YOSHINORI OTSUKI AND YUKO ITO

Introduction

Transmission electron microscopy (TEM) gives us much morphological information on apoptosis, since the concept of apoptosis proposed by Kerr et al. (1972) was based on several characteristic ultrastructural findings. Agarose gel electrophoresis using extracted DNA is preferentially used in the molecular-biological studies dealing with apoptosis and reveals the apoptotic nuclear changes such as nucleosomal DNA fragmentation i.e., DNA ladder (Arends et al. 1990). In addition to TEM and agarose gel electrophoresis, in situ end labeling of DNA strand breaks (ISEL) technique using light microscopy (LM)(conventional ISEL technique) which labels free 3'-OH ends is widely used for the discrimination between apoptotic cells and necrotic cells. Recently, much interest has been focused on the problems related to the specificity of both the agarose gel electrophoresis of extracted DNA and the conventional ISEL technique in the detection of cell death, although both are far more convenient methods than TEM.

- Agarose gel electrophoresis can not demonstrate the DNA ladder when the sample contains only a few apoptotic cells (Herrmann et al. 1994)

Correspondence to Yoshinori Otsuki, Osaka Medical College, Department of Anatomy and Biology, 2-7, Daigaku-machi, Takatsuki, Osaka, 569, Japan (*phone* +81-726-831221; *fax* +81-726-846511; *e-mail* an1001@art.osaka-med. ac. jp); Yuko Ito, Osaka Medical College, Department of Anatomy and Biology, 2-7, Daigaku-machi, Takatsuki, Osaka, 569-8686, Japan

- The conventional ISEL technique is positive not only for apoptotic cells but also for necrotic cells (Ansari et al. 1993; Wijsman et al. 1993; Inoki et al. 1997).
- The ISEL using TEM (ISEL/TEM) technique offers more considerable advantages than the conventional ISEL technique and agarose gel electrophoresis.
- When the sample contains only a few apoptotic cells, the ISEL/TEM technique enables to not only discriminate the apoptotic cells among many cells but also to demonstrate the ultrastructural characteristics of apoptosis.
- Necrotic cells labeled with the ISEL/TEM technique are easily differenciated from the labeled apoptoic cells due to the ultrastructural characteristics between them.
- The ISEL/TEM technique combined with image analysis can semi-quantify the sites of free 3'-OH DNA ends in apoptotic cells, when immunogold particles are used in the ISEL/TEM technique, as a substitution for the immunoperoxidase in the conventional ISEL technique.

Outline

The entire protocol of the ISEL/TEM technique is outlined in Fig. 1.

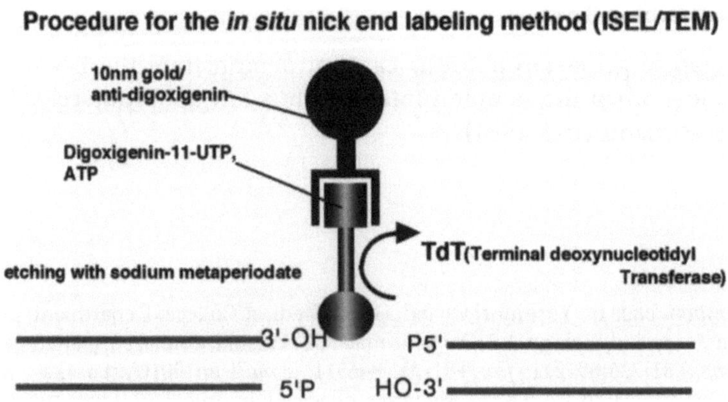

Fig. 1. The entire protocol of the ISEL/TEM technique as a simplified schema.

Materials

Equipment

- computer (e.g., Power Macintosh 7100/66AV, Apple Computer Inc., California, USA)
- scanner (e.g., Epson GT-6000, Epson Inc., Nagano, Japan)
- softs (e.g., Adove Photoshop™, Adove Systems Incorporated, California, U.S.A.; NIH Image program, written by Wayne Rasband at the U.S. National Institutes of Health and available from the Internet by anonymous ftp from zippy. nimh. nih. gov or on floppy disk from NTIS, 5285 Port Royal RD., Springfield, VA 22161, part number PB 93-504868)
- ultramicrotome (e.g., Sorval® MT-6000, Du Pont Company, Delaware, U.S.A.)
- transmission electron microscope (e.g., H-7100, Hitachi, Tokyo, Japan)
- incubator (e.g., ESPAC LN-122, TABAI Corp., Tokyo, Japan)

Tools

See also Fig. 2.

- nickel grids (150 meshes)
- locking forceps
- anticapillary forceps

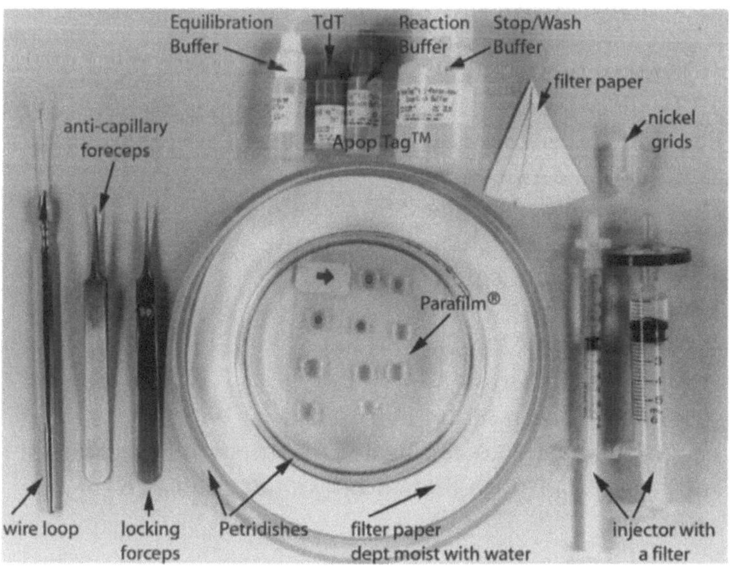

Fig. 2. The tools and reagents used in the ISEL/TEM technique.

- wire loop
- injector with a filter (0.2 μm in pore size)
- Petri dishes
- Parafilm® (American National Can™, Connecticut, U.S.A.)
- filter paper

Reagents

- Fixative
  - 0.1M Phosphate buffer (PB) (pH7.4)
  - 2% paraformaldehyde
  - 2.5% glutaraldehyde
- Post-fixative
  - 1% osmium tetroxide in 0.1 M PB
- Epoxy resin (Total volume 20 ml)

Epon 812	9.41 ml
MNA	5.65 ml
DDSA	4.94 ml
DMP-30	0.4 ml

- saturated sodium metaperiodate: 1g sodium metaperiodate in 8ml distilled water
- Equilibration Buffer for the ISEL technique
  - 30 mM Tris-HCl (pH 7.2)
  - 140 mM sodium cacodylate
  - 1 mM cobalt chloride
- 0.3 U/ μl TdT for the ISEL technique
- Reaction Buffer for the ISEL technique: 0.04 nmol digoxigenin-dUTP, dATP
- Stop/Wash Buffer for the ISEL technique
  - 300 mM sodium chloride
  - 30 mM sodium citrate
- Tris-BSA buffer
  - 0.02 M Tris-HCl buffer (pH7.4)
  - 0.1% Bovine Serum Albumin
- 10% normal sheep serum in Tris-BSA
- 10 nm colloidal gold conjugated to sheep anti-digoxigenin antibody (e.g., British BioCell Int, Golden Gate, UK)

Procedure

Preparation of samples

1. Cut the specimens into small pieces (5 X 5 X 5 mm in maximal size).
2. Fix with the fixative for 2 hrs at 4 °C.
3. Rinse and immerse in PB overnight at 4 °C.
4. Post-fix with osmium tetroxide for 2 hrs at room temperature (RT).
5. Immerse in 0.1 M PB for 5 mins, three times.
6. Dehydrate in a series of graded ethanol solutions for 15 mins, respectively.
7. Clear in propylene oxide for 15 mins, twice.
8. Embed in epoxy resin.
9. Make ultra-thin sections (more than 80 nm thick) by an ultramicrotome and mount on nickel grids.

ISEL/TEM technique

Several buffers for the ISEL/TEM technique shown in the Materials Section are cited from the method of TdT-mediated dUTP-biotin nick end labeling (Gavrieli et al 1992). We recommend the use of an available commercial non-radioactive apoptosis detection kit (e.g., Apop Tag™ : Oncor, Gaithersburg, MD, USA) to obtain constant and trustworthly data, although the purchase of the kit may be expensive. Therefore, we will state here the procedure for the ISEL/TEM technique, using Apop Tag™ kit (see Fig. 2).

Each reaction is performed on droplets of medium placed on Parafilm; the face of the grid with ultra-thin sections is floated, incubated or washed on the droplets of the medium (Fig. 3).

Know how to use locking forceps, anticapillary forceps and wire loop. We use locking forceps to securely grip the grid, while washing in the beaker. Anti-capillary forceps are used when the grid is transported between the droplets consisting the different

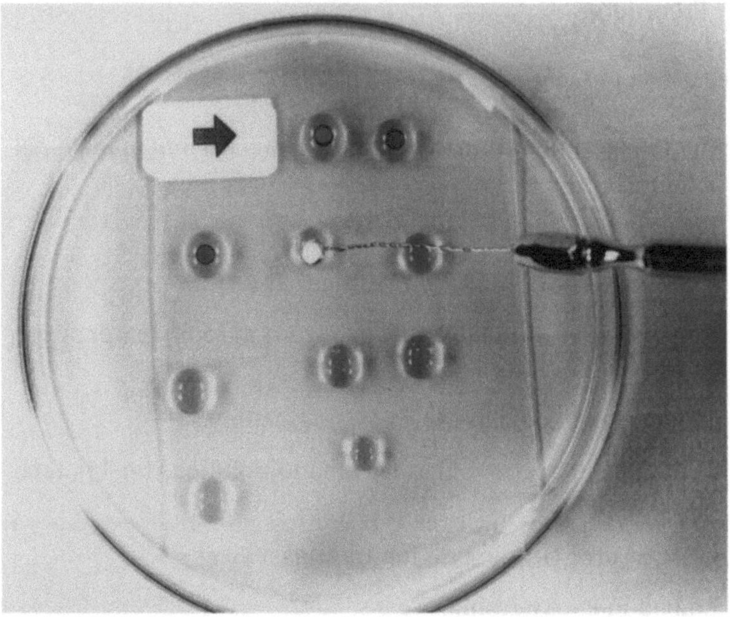

Fig. 3. The grid floating on the droplet is recovered using the wire loop and is transported to the droplet consisting of the same medium.

medium. The wire loop is used when the grid is transported between the droplets consisting of the same medium.

Note: After the incubation with the medium, carefully blot the excessive medium both at the tip of the anti-capillary forceps and on the face of the grid, using small pieces of filter paper.

1. Etch the ultra-thin sections with saturated sodium metaperiodate for 2 mins at RT.
2. Rinse in distilled water, for 5 mins, twice.
3. Incubate in the Equilibration buffer (Apop Tag^{TM}), for 10 to 15 secs.
4. Incubate in the mixture of Reaction buffer and TdT (Apop Tag^{TM}) for 60 mins, at 37 °C.
5. Incubate in the Stop/wash buffer (Apop Tag^{TM}), for 30 mins at 37 °C.
6. Rinse in Tris-BSA buffer, 5 mins, three times.

7. Incubate in normal sheep serum, for 15 mins.
8. Incubate in the anti-digoxigenin antibody conjugated to colloidal gold diluted to 1 : 50 with Tris-BSA for 60 mins.
9. Rinse in Tris-BSA buffer for 5 mins.
10. Rinse in distilled water, for 5 mins, three times.
11. Stain with uranyl acetate for 5 mins and then stain with lead citrate for 1 min.
12. Wash with distilled water in a beaker.

Image analysis via NIH image program

1. Observe the ultra-thin sections using a TEM and take photographs of the labeled cell at high magnification (X 15,000 to 20,000X) (Fig. 4).
2. Import the images of the whole cell from the electron micrographs on a computer via a scanner.
3. Leave only the image of the nucleus on a computer using a soft <Adove Photoshop™> (Fig. 5).
4. Automatically measure the nuclear area (μm^2) and count the number of the immunogold particles via NIH Image program (Fig. 6).
5. Calculate the labeling density for DNA strand breaks (the number of immunogold particles / μm^2).

Results

In the conventional ISEL technique, the reaction products of DAB oxidized by horseradish peroxidase demonstrate the localization of 3'-OH DNA ends in apoptotic cells as a brown color (DAB reaction).

- Melanosomes of epidermal cells and granules of eosinophils with peroxidase activity show a false positive 3, 3'-diaminobenzidine (DAB) reaction when they are incubated with a solution containing DAB (Fig. 7). Therefore, the conventional

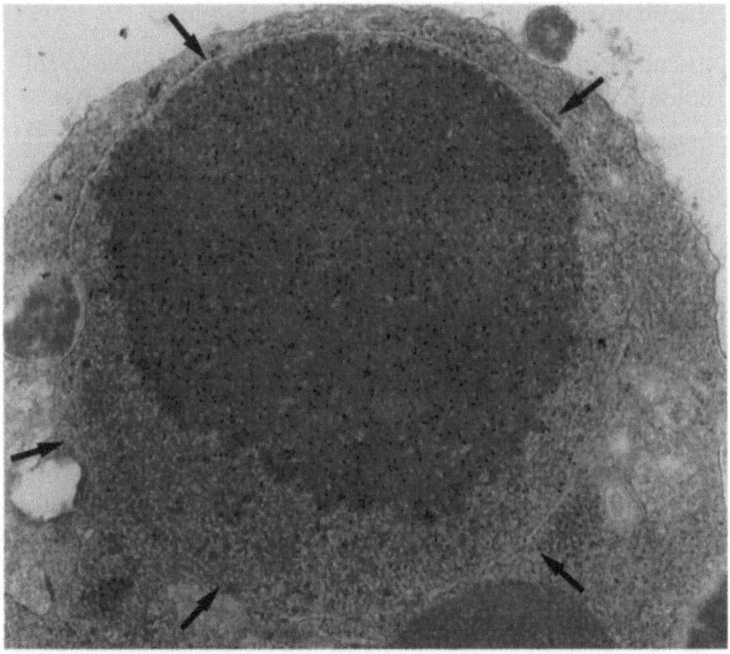

Fig. 4. Electron micrograph stained with the ISEL/TEM technique (X 15,000). The nuclei (arrows) of apoptotic endometrial glandular cell at the late secretory phase is characterized by the presence of condensed chromatin with many immunogold particles. **Note** the presence of many immunogolds (shown in black spots) only at the condensed chromatin but not at the other nuclear chromatin.

ISEL technique involving immunoperoxidase method is unsuitable for the study of the epidermis and eosinophils.

The target of the ISEL/TEM technique is the newly formed 3'-OH DNA ends. A few immunogold particles indicating 3'-OH DNA ends are observed on the evenly distributed chromatin of cells undergoing proliferation, although the conventional ISEL technique does not stain the proliferous cells. This finding may be due to an artifact from oxidant treatment, section cutting (Ansari et al. 1993), DNA breaks induced by DNA repair systems or sites of active gene transcription located in the chromatin (Thiry 1991), rather than oligonucleosomal degradation.

- The labeling is certainly conditioned by the concentration and incubation time of oxidants for etching. The application of saturated sodium metaperiodate for only 2 mins yields

Fig. 5. The nucleus shown in Fig. 4 is left on the computer using a soft <Adove Photoshop^{TM}.

good results in terms of the background and specificity (Inoki et al. 1997).

If you use the ISEL/TEM technique in your study, you will easily understand that cells labeled with the conventional ISEL technique contain various stages of apoptotic cells; cells with increased perinuclear chromatin but otherwise with normal ultrastructural appearance, cells with fragmented nuclei in the cytoplsm and apoptotic bodies.

- Free 3'-OH DNA ends can be detected in the relative early apoptotic cells via the ISEL/TEM technique, prior to the ultrastructural changes of apoptosis.

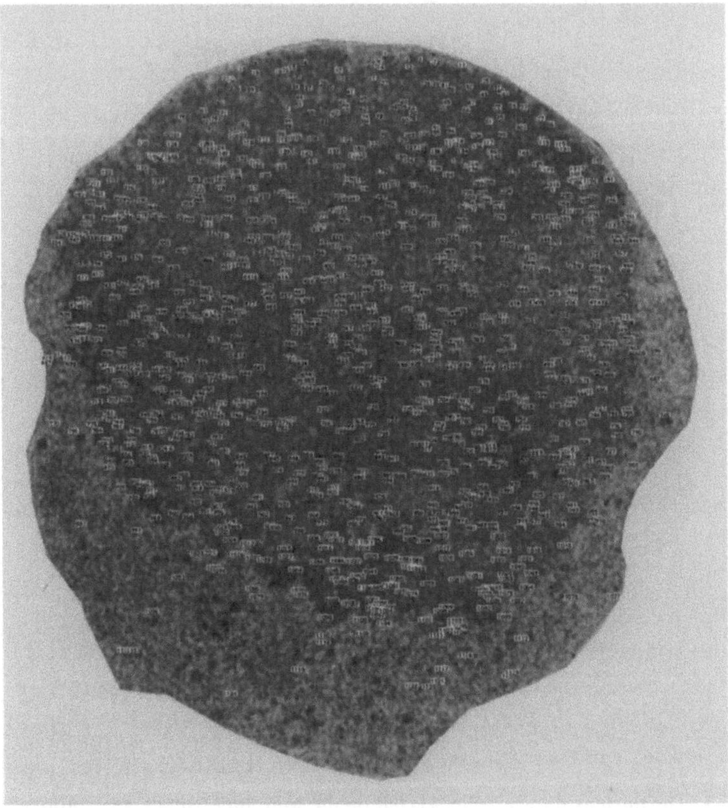

Fig. 6. The number of immunogolds and nuclear area are automatically measured via the NIH Image program.

Moreover, you will notice the various ultrastructural changes during apoptotic process, if you use several different cultured cells or tissues, and even if using the same cell line treated with different stimuli such as heat shock and anti-cancer drugs (Fig. 8).

The conventional ISEL technique stains not only apoptotic cells but necrotic cells. The ISEL/TEM technique clearly demonstrates many immunogold particles indicating the sites of 3'-OH ends in the nuclei of necrotic cells (Fig. 9) and supports the above finding of the conventional ISEL technique.

When you use image analysis combined with the ISEL/TEM technique, you can semi-quantify DNA strand breaks in cells undergoing apoptosis and necrosis. The image analysis for quan-

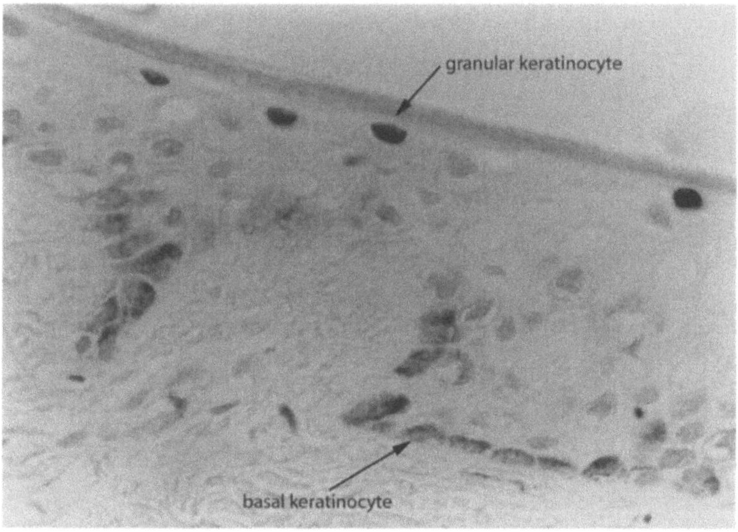

Fig. 7. Human epidermis stained with the conventional ISEL technique. **Note** the presence of the positive reaction in the granular keratinocytes and of the false positive one in the basal keratinocytes. X400

titative changes of DNA strand breaks is performed on a computer using the NIH Image program which enables the measurement of the nuclear area (μm^2) and the number of DNA strand breaks labeled with immunogold particles on the nucleus. The labeling density for DNA strand breaks (the number of immunogold particles / μm^2 is evaluated in apoptotic cells of the several stages (Inoki et al. 1997) (Fig. 10).

Troubleshooting

- Only a few immunogold particles were detected on the labeled cells:
  - The face of grid without ultra-thin sections may have been floated and incubated on the droplets of the medium.
  - DNA volume contained in the ultra-thin section may be scanty; check the thickness of the sections. Make thicker ultra-thin sections (at least 80 nm thick is recommended).
  - Epoxy resin covering free 3'-OH DNA ends may not be sufficiently etched by saturated sodium metaperiodate; check the time for etching (usually only 2 mins is optimal time).

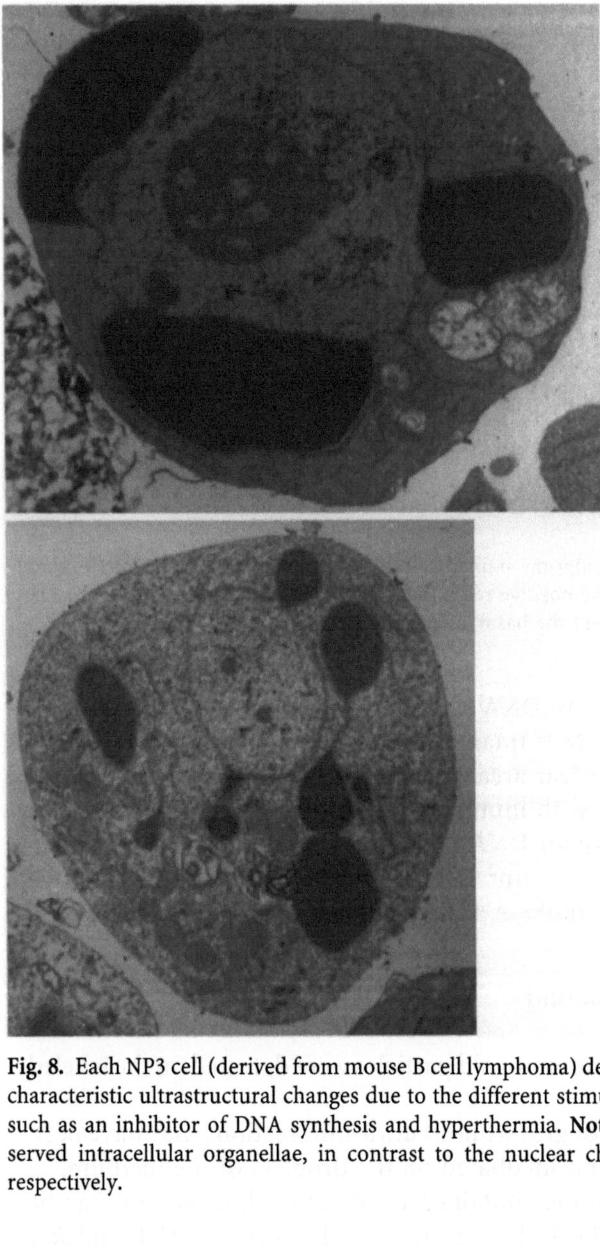

Fig. 8. Each NP3 cell (derived from mouse B cell lymphoma) demonstrates the characteristic ultrastructural changes due to the different stimuli of apoptosis such as an inhibitor of DNA synthesis and hyperthermia. **Note** the well preserved intracellular organellae, in contrast to the nuclear changes. X5,000, respectively.

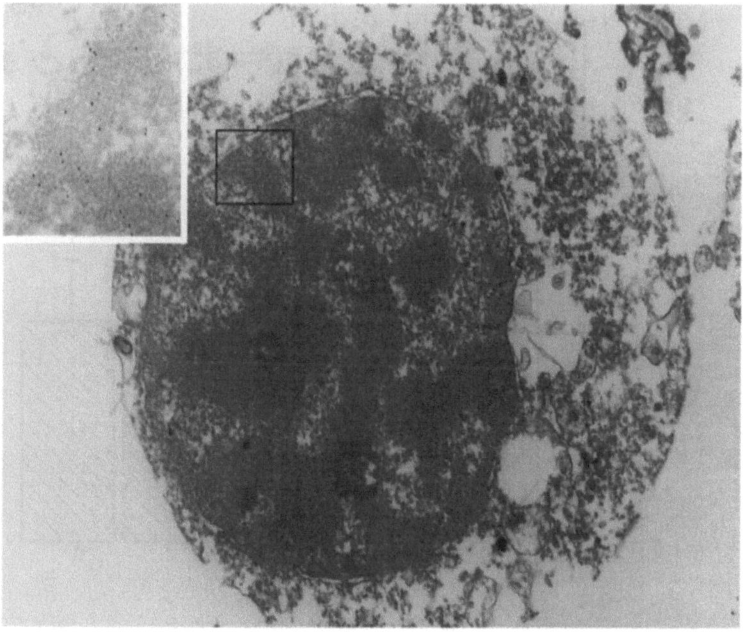

Fig. 9. A necrotic cell labeled with the ISEL/TEM technique. The many immunogold particles are observed on the agglutinated chromatin showing the fine granular pattern. X 12,000. The inset (X20,000) corresponds to the boxed area.

  - Check the optimal dilution of the anti-digoxigenin antibody conjugated to colloidal gold.

- Immunogold particles are present throughout the ultra-thin section, as background:
  - The grids may have sunk at the bottom of the droplets on the way, throughout the staining process; carefully blot the excessive medium both at the tip of the forceps and on the face of the grid, using small pieces of filter paper.
  - Do not dry the sections during the staining process.

- Immunogold particles present on the chromatin or chromatin condensation are not counted via NIH Image program:
  - Heavy metals such as gold and silver show the most intense electron-density among substances on the electron micrographs. The NIH Image program recognizes and counts the immunogolds based on the difference of the contrast be-

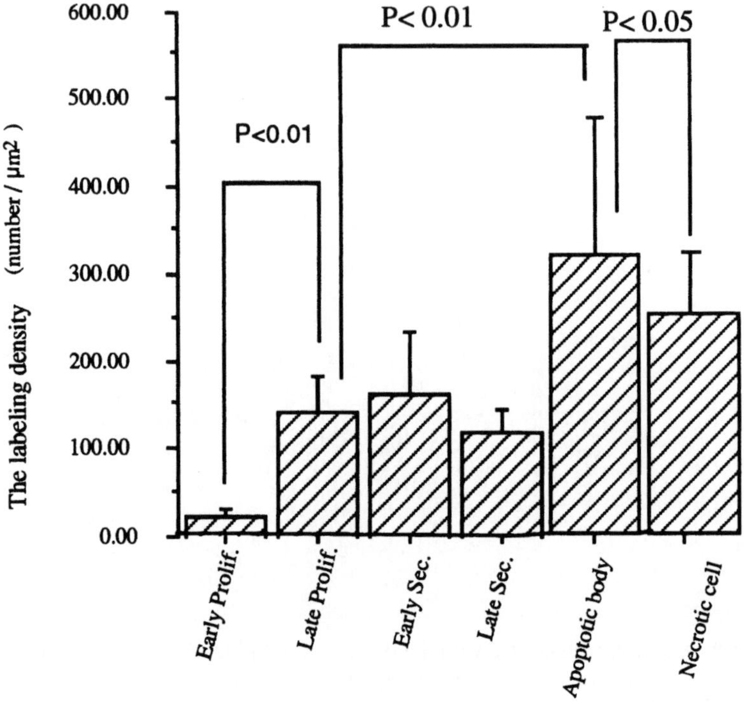

Fig. 10. The statistical analysis for the labeling density of free 3'-OH DNA ends in the uterine glandular cells in each menstrual phase, apoptotic cells and necrotic cells.

tween immunogolds and other substances; omit the staining step of lead citrate.

- Contamination showing high electron density is counted via NIH Image program:
  - Remove the contaminations from the images using a soft <Adove Photoshop™ >prior to the process of NIH Image program.

Applications

The ISEL/TEM technique is applied for confocal laser microscopy (CLM). The site of feee 3'-OH DNA ends is detected by the reflectance from heavy metal products (anti-digoxigenin antibody-immunogold-silver complexes), instead of DAB reaction products in the conventional ISEL technique or immunogold complexes in the ISEL/TEM technique. Methyl green has been known to stain double-stranded DNA (Burres et al. 1993; Melnick and Pickering 1988) and is weakly autofluorescent (Rothbarth et al 1976). The ISEL using CLM technique combined with methyl green staining enables the simultaneous localization of both 3'-OH DNA ends and double-stranded DNA in apoptotic cells and also, the comparison of their concentrations (Ito and Otsuki 1998).

References

Ansari B, Coates PJ, Greenstein BD, Hall PA (1993) *In situ* end-labeling detects DNA strand breaks in apoptosis and other physiological and pathological states. J Pathol 170:1-8

Arends MJ, Morris RG, Wyllie AH (1990) Apoptosis: the role of the endonuclease. Am J Pathol 136:593-608

Burres NS, Frigo A, Rasmussen RR, McAlpine JB (1992) A colorimetric microassay for the detection of agents that interact with DNA. J Nat Prod 55:1582-1587

Gavrieli Y, Sherman Y, Ben-SAsson SA (1992) Identification of programmed cell death in situ via specific labeling of nuclear DNA fragmentation. J Cell Biol 119:493-501

Herrmann M, Lorenz HM, Voll R, Grünke M, Woith W, Kalden JR (1994) A rapid and simple method for the isolation of apoptotic DNA fragments. Nucleic Acids Research 22:5506-5507

Inoki C, Ito Y, Yamashita H, Ueki K, Fukuda Y, Ninomiya E, Nakamura K, Hayashi R, Ueki M, Otsuki Y (1997) Image analysis and ultrastructural detection of DNA strand breaks in human endometrium by *in situ* end-labeling techniques. J Histotechnol 20:321-328

Ito Y, Otsuki Y (1998) Localization of apoptotic cells in the human epidermis by an *in situ* DNA nick end labeling method using confocal reflectant laser microscopy. J Histochem Cytochem (in press)

Kerr JFR, Wyllie AH, Currie AR(1972) Apoptosis: a basic biological phenomenon with wide-ranging implications in tissue kinetics. Br J Cancer 26:239-257

Melnick J, Pickering M (1988) The mechanism of the drug induced partial displacement of methyl green from DNA. Biochem Int 16:69-75

Rothbarth PH, Hendriks-Sturkenboom I, Ploem JS (1976) Identification of monocytes in suspensions of mononuclear cells. Blood 48:139-147

Thiry M (1991) *In situ* nick translation at the electron microscopic level: a tool for studying the location of DNase-sensitive regions within the cell. J Histochem Cytochem 39:871-874

Wijsman JH, Jonker RR, Keijzer R,van de Velde CJH, Cornelisse CJ,van Dierendonck JH (1993) A new method to detect apoptosis in paraffin sections: in situ end-labeling of fragmented DNA. J Histochem Cytochem 41:7-12

Abbreviations

CLM	confocal laser microscopy
DAB	3, 3'-diaminobenzidine
LM	light microscopy
ISEL	*in situ* nick-end labeling
ISEL/LM technique	ISEL technique using light microscopy
ISEL/TEM technique	ISEL technique using transmission electron microscopy
RT	room temperature

Part II

Chapter 5

Introduction to the Detection of Specific DNA and RNA Sequences

TAKEHIKO KOJI

Historical remarks

In situ hybridization was originally developed by Gall and Pardue (1969) and John et al. (1969) as a method to localize specific DNA sequences in cytological preparation. Afterwards, the method was used as a powerful technique to detect the locus of a specific gene in chromosomes or the sites of viral DNA. In the late 1970s, Brahic and Haase (1978) revealed that the same principle is applicable to localize specific RNA in cell and tissue preparations. In keeping pace with the burst prosperity of molecular cloning fields, in situ hybridization has progressed and now has turned out to be one of the most essential techniques in cell biology and developmental biology. In this chapter, I will try to overview the methodological aspects of in situ hybridization, particularly focusing on the detection of specific RNA in tissue sections, in which we have been actively engaged for more than 15 years (Koji and Nakane 1990, 1996).

For a better understanding of the regulatory mechanisms of cell-type or tissue-type specific gene expression, it is essential to analyze the expression of specific genes at the level of mRNA as well as protein in individual cells. To localize proteins at a cellular level, immunohistochemistry has been a powerful tool. However, the information obtained by immunohistochemistry on intracellular protein synthesis was limited because the protein might be secreted, degradated, or post-translationally modified. Also, it was known that the best performance of immunos-

Correspondence to Takehiko Koji, Nagasaki University School of Medicine, Department of Histology and Cell Biology, 1-12-4 Sakamoto, Nagasaki, 852-8523, Japan (*phone* +81-095-849-7025; *fax* +81-095-849-7028; *e-mail* tkoji@net.nagasaki-u.ac.jp)

taining requires optimization of the protocol depending upon the stability of an epitope and the specimens which were not processed appropriately for each antigen or antibody to be investigated could result in false-negative or underestimated findings (Ruizeveld de Winter et al. 1991; Slayden et al. 1995; Ehara et al. 1996). Practically, however, the optimal conditions cannot always be guaranteed for immunostaining with an antibody. Thus, the use of in situ hybridization for mRNA has been indispensable not only for assessment of the expression of specific mRNA, but also for verification of in situ synthesis of specific proteins (Koji and Nakane 1996).

In situ hybridization depends upon base-pairing between target nucleic acids in cells or tissues and probe nucleic acids and its conditions have been already optimized. The probe must be labeled with either radioactive or nonradioactive molecules to reveal the sites of hybridization. Originally, the probe nucleic acids were labeled with radioisotopes such as ^{3}H, ^{35}S, ^{32}P, ^{33}P and ^{125}I and the sites of hybrids were visualized by autoradiography. As shown in Table 1, however, the use of a nonradioactive probe is considered superior to a radioactive one from viewpoints of resolution, probe stability, a shorter processing time and personal safety.

A variety of nonradioactive labels were so far employed and most of them were haptenic compounds such as sulfone, biotin

Table 1. Comparison of Radio and Hapten Labeled Probes

	Autoradiography	Immunohistochemistry
Half-life	^{32}P (14 d) ^{35}S (88 d) ^{3}H (12 y)	Infinite
Time for processing	1 - several weeks	1 - 2.5 d
Resolution	Poor (tissue level)	Excellent (cell level)
Sensitivity	High	High
Quantification	Possible	Possible
Double localization	Difficult	Easy
Application for EM	Difficult	Easy
Personal safety	Bad	Good

and digoxigenin (Dig). Among various nonradioactive labels, thymine-thymine (T-T) dimer method developed by us previously (Koji and Nakane 1990) is unique because T-T dimer can be easily introduced by ultraviolet light (UV) irradiation on a DNA solution and it does not require any purification of labeled probe from unreacted compounds. In this chapter, I will use T-T dimer method as well as Dig method (Koji and Brenner 1993; Koji et al. 1994) for example, because of the close sensitivity between them. The overview will include recent progress in optimization of the protocol and enhancement of signal intensity.

Principle of in situ hybridization

DNA consists of 4 bases, i.e. adenine (A), guanine (G), cytosine (C) and thymine (T) and in the case of RNA thymine is replaced with uracil (U). A can form a stable complex with T or U through two hydrogen bonds and G can bind to C through three hydrogen bonds, as shown in Fig. 1. Therefore, A to T or U and G to C, vice versa, are regarded as complementary bases. This is a basic principle for nucleic acid hybridization. If a base sequence on a single nucleic acid strand is complementary to that of the other strand, they can form stable molecular hybrids to be double-stranded.

The stability of molecular hybrids highly depends upon temperature, a base sequence and a salt concentration. As a physicochemical parameter of the stability, Tm (melting temperature) is often used, showing a temperature where a half of bases are complex, if simplified. Tm between DNA and DNA can be calculated by the following empirical equations; Tm=81.5 + 16.6 log[Na+]+0.41(%GC)-820/bases-0.61(%F) (for a DNA probe with more than 100 bases) (Thomas et al. 1973; Meikoth et al. 1984), Tm=81.5 + 16.6 log[Na+]+0.41(%GC)-675/bases-0.61(%F) (for an oligodeoxynucleotide (oligo-DNA) probe) (Henderson et al. 1991). Here, [Na+] ; a concentration of salt in the reaction mixture, %GC; total number of G and C/total number of bases, bases; total number of bases in the probe, %F; a concentration of formamide, which is thought to work as a hydrogen-bond blocker like urea and decreases the Tm value. When there are some base mismatches between the probe and target sequences, 1.4 x (%mismatch) should be subtracted from Tm. Tm between RNA and RNA is higher by 10-15 °C than that of Tm between DNA and DNA in the same base composition, and Tm in DNA-RNA hybrids is supposed to be a halfway between Tm in DNA-DNA hybrids and Tm in RNA-RNA hybrids.

The parameters influencing the rate of hybrid formation were already well established in liquid phase (Wetmur et al. 1968). However, only a little information on the hybridization reaction between solid phase and liquid phase, which will be much more important for in situ hybridization, can be obtained. Collectively, the rate of hybrid formation is dependent upon temperature of hybridization and a concentration of probe, while the rate of hybrid formation is not so changed at the salt concentration of 0.4-1.0 M and at pH of 5.0-9.2. We usually perform in situ hybridization at (Tm-25) °C for longer probes and (Tm-15) °C for oligo-DNA probes with 30-45 bases.

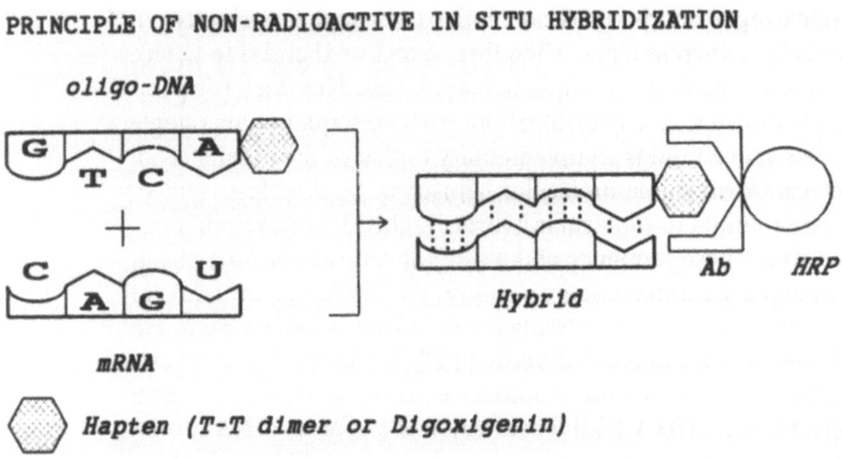

Fig. 1. Principle of nonradioactive in situ hybridization with an oligo-DNA probe. (1) T-T dimerized or Dig-labeled oligo-DNA is hybridized in situ with a tissue section. (2) After appropriate washing, T-T dimer or Dig is reacted with horseradish peroxidase (HRP)-labeled anti-(T-T dimer) or anti-Dig. (3) Finally, the sites of HRP are visualized enzyme-immunohistochemically.

Basic factors affecting the signal detection

During our efforts to establish the nonradioactive in situ hybridization protocol for specific RNA detection, we realized that there are several basic points to be optimized. Some of them are related to probe nucleic acids and the others are associated with specimens.

Check points for probes

- As probes for in situ hybridization, there are several choices; double-stranded cDNA, single-stranded cDNA, oligo-DNA and cRNA. Depending upon the selection of probe nucleic acid, what kind of labels can be chosen and how the probe nucleic acid is labeled are different. Irrespective of selected labels or of labeling protocols, it is essential for successful implementation of nonradioactive in situ hybridization to verify whether the probes are optimally labeled. Too little labeling results in a low signal, and excessive labeling will reduce the efficiency of hybridization and increase the risk to be suffered from the label-compound specific background (e.g., an excess introduction of Dig to oligo-DNA results in an increase in the hydrophobicity of probe and becomes sticky to any hydrophobic parts of tissue sections). These issues can be mostly assessed by dot-blot hybridization(Koji and Nakane 1990), as shown in the Chart "Procedures for dot-blot hybridization with T-T dimerized oligo-DNA probe". Generally speaking, when you get 1 pg-10 pg of complementary DNA fixed onto a nitrocellulose or nylon membrane by the colorimetric staining, your probe will be expected to work in in situ hybridization as well.

- The other aspect that a probe should fulfil if used for in situ hybridization is that the labeled probes must be able to penetrate into cells and tissue sections. It has been observed frequently when large-size nucleic acids (more than 500 bases) are haptenized whether chemically or physically and no nuclease is used during the process of haptenization, the length of the DNA stays the same and is too large to penetrate into cell or tissue sections. The best size of probe nucleic acid varies according to the conditions of fixation (Moench et al. 1985) and the degree of protease digestion during pretreatment (Koji et al. 1988). With routine pretreatment protocols, a size of less than 200-400 bp (as assessed by ethidium bromide staining of agarose-gel after electrophoresis of probe DNA) is recommended. Since the usual size of oligo-DNA is less than 100 bases, no significant problem is encountered.

Procedures for dot-blot hybridization with T-T dimerized oligo-DNA probe

1. Cut nitrocellulose filters and immerse them in 20 X SSC (1 X SSC=0.15 M NaCl/ 0.015 M Na citrate (pH 7.0)) for 30 mins, and air-dry on Parafilm.
2. Apply sense oligo-DNA (boiled for 5 mins and chilled quickly at 4 °C) to the filters at 1 pg to 10 ng/2 µl/spot.
3. After air-drying, bake the filters for 2 hrs at 80 °C.
4. Prehybridize for 2 hrs at 37 °C in a medium (boiled for 7 mins and chilled quickly at 4 °C) consisting of 10 mM Tris/HCl buffer (pH 7.4) containing 0.6 M NaCl, 1 mM EDTA, 1 X Denhardt's solution, 500 µg/ml yeast tRNA, 250 µg/ml salmon testis DNA and 40% formamide.
5. Hybridize at 37 °C overnight in a medium (boiled for 7 mins and chilled quickly at 4 °C) consisting of 10 mM Tris/HCl buffer (pH 7.4) containing 0.6 M NaCl, 1 mM EDTA, 1 X Denhardt's solution, 250 µg/ml yeast tRNA, 125 µg/ml salmon testis DNA, 10% dextran sulfate, 40% formamide and 1-2 µg/ml T-T dimerized antisense oligo-DNA.
6. Wash with 2 X SSC (30 mins, 2 times), 1 X SSC (30 mins, 2 times), 0.5 X SSC (at room temperature to 45 °C; 15 mins, once) and PBS (15 mins, once), successively.
7. Immerse for 1 hr in PBS containing 5% BSA, 500 µg/ml normal mouse IgG, 100 µg/ml yeast tRNA and 100 µg/ml salmon testis DNA.
8. React for 1 hr with HRP-labeled monoclonal anti-(T-T dimer) antibody dissolved in PBS containing 5% BSA, 100 µg/ml yeast tRNA and 100 µg/ml salmon testis DNA.
9. Wash 5 times with 0.075% Brij in PBS for 15 mins each time.

10. Wash once with PBS for 15 mins.

11. Visualize the HRP sites in a chromogen solution.

12. Wash 3 times with deionized, distilled water for 15 mins each time, and dry up.

Check points for cell or tissue preparation

- We can use any kinds of specimens for in situ hybridization, including cultured cells (Koji et al. 1989), fresh frozen tissues (Koji and Brenner 1993), pre-fixed tissues with some fixative (Goto et al. 1990; Ueki et al. 1995), paraffin-embedded tissues (Razzaque et al. 1994; Hakuno et al. 1996) and resin-embedded tissues (Nakane et al. 1990). The most important point in specimens which will be used for in situ hybridization of specific mRNA is to assess the integrity and the hybridizability of RNA.

The methods to stain RNA with dyes in tissue sections or cells are known; one is done with acridine orange and the other one with methyl green/pyronin Y. Both staining methods are quantitatively reliable for the assessment of total RNA. However, they do not provide any information on the accessibility of the RNA to probes. In fact, intense RNA staining with these dyes in specimens does not always mean its high hybridizability, and the discrepancy between staining intensity and hybridizability is very common in specimens fixed well with cross-linking fixatives. Again, although it has been widely accepted that the degree of proteinase K digestion in the pretreatment step profoundly influences the signal intensity, especially in paraffin sections of formalin-fixed tissues, it had essentially no effect on the total RNA staining with methyl green/pyronin Y (Yoshii et al. 1995).

For assessment of hybridizability of tissue RNA, in situ hybridization of widely distributed and highly conserved RNA seemed to be preferable. Consequently, we chose ribosomal RNA (rRNA) as a candidate. rRNA is thought to be a superior reference to β-actin mRNA and poly(A) because an occasional failure to detect the β-actin mRNA may not be due to a loss of RNA from specimens. Rather it may be due to an extreme paucity of the mRNA, and the probe to poly(A) does not contain G and C, resulting in the unusual low Tm value of hybrids. By computer-assisted searching of rat 28S rRNA cDNA, a segment consisting of 34 nucleotides, which are conserved beyond species (ie, 100% homology of the sequence was found in mouse, rat and human), was selected and the oligo-DNA complementary to the segment of 28S rRNA with auxiliary ATT repeats at the 3'- and 5'- ends was synthesized (Yoshii et al. 1995), as follows; 5'-TTATTA-TGCTACTACCACCAAGATCTGCACCT GCGGCGGC ATTAT-TATT-3'. Here, ATT repeats were added to be haptenized with T-T dimer formed by UV irradiation.

With this probe, 28S rRNA was clearly localized in the cytoplasmic area as well as in nucleoli in various cells and tissue sections and we believe that 28S rRNA staining by in situ hybridization is a convenient parameter for evaluation of levels of hybridizable RNAs in tissue sections. Also, we confirmed the effectiveness of 28S rRNA staining as reference for optimization of the pretreatment conditions including protease digestion.

Essential steps in the pretreatment

Fixation To immobilize mRNA and to preserve the cell and tissue morphology, fixation is an essential step. As a fixative for in situ hybridization, both protein- precipitating fixatives such as ethanol/acetic acid (3 : 1) and protein cross-linking fixatives such as formaldehyde and glutaraldehyde have been used (Koji and Nakane 1990). It should be noted that although stronger fixation (e.g., a longer exposure of specimens to paraformaldehyde or fixation with glutaraldehyde) results in higher levels of the retention of cellular RNA, it is known to decrease the penetrability of probe nucleic acids into cells or tissue sections (Moench et al. 1985; Lawrence and Singer 1985). In our hands, the most favorable fixation is obtained with 4% paraformaldehyde in PBS (pH 7.4) for in situ hybridization of specific RNA. However, we do not care so much even if we encounter specimens processed under suboptimal conditions, because we are routinely assessing the retention of hybridizable RNA, as described above.

Protease treatment In fixed materials, mRNA is supposed to be embedded in cellular matrix proteins and prevented from forming hybrids with probe nucleic acids. Consequently, unmasking of mRNAs by removing proteins from the specimens is often required for full implementation of in situ hybridization. To accomplish this task, the most striking method is the use of proteolytic enzymes such as proteinase K, pronase and pepsin. With any protease, optimization of the proteolysis conditions (concentration, digestion time, temperature, etc.) for a given tissue is prerequisite. Over-digestion is severely deteriorative to the morphology as well as the retention of RNA. Generally, tissues fixed with cross-linking fixatives absolutely require protease treatment, whereas those fixed with protein-precipitating fixatives do not always need this step. The optimal conditions of proteolysis also depend upon the length of probe nucleic acids; longer probes require stronger treatment (Koji et al. 1988). Even when a short DNA (less than 50-60 bases) such as oligo-DNA is employed, protease digestion is still often effective (Koji et al. 1989; Yoshii et al. 1995), indicating that the protease unmasks or exposes the target mRNAs as well as enhancing the penetration of the probe. In addition, the necessity of protease digestion is also different depending upon what kind of preparation or what kind of tissue is used.

Control experiments

To verify the specificity of the signals obtained by in situ hybridization, various controls are necessary. The controls usually performed are as follows:

(a) Pretreatment with RNase and DNase. If the signals are derived from hybridization with RNA, the signals must be lost after RNase treatment, while DNase does not affect the signals.

(b) Use of control probes. If probes bearing sequences unrelated to the target mRNA produce no significant staining, the signals obtained with the proper probe must be the result of specific hybridization. In the case of single-stranded probes, sense strand probe is conventionally used as a control probe. When synthetic oligo-DNA probes are used, nonsense probe, which has the same base composition to the antisense probe, but with a different sequence, will be employed.

(c) Blocking or competition experiment. If an excess amount of non-labeled probe nucleic acid complementary to target mRNA is preincubated or added to the hybridization mixture and the signal is abolished or markedly decreased, the signal is considered to be specific. In the blocking control, we should show that non-labeled unrelated nucleic acid cannot block the successive hybridization with the proper probe. In the competition experiment, it is also essential to show that no changes occur in the presence of an excess amount of non-labeled unrelated nucleic acid, neglecting the effect of simple dilution of the proper probe.

(d) Excessively high stringent washing. When the cells or tissue sections, which were already hybridized with a specific probe, are washed at a higher temperature than the expected Tm value of the hybrids, the signal should be dislodged.

(e) Biological control. No positive signal is found in specimens which are known to lack the specific target mRNA.

(f) If possible, it should be demonstrated that the target tissue expresses only one species of mRNA which hybridizes with the probe by Northern blotting.

I will show an example of a set of control experiments in Fig. 2, where estrogen receptor mRNA was localized in situ with Dig-labeled antisense oligo-DNA probe in the fresh frozen sections of monkey uterus (Koji and Brenner 1993).

New approaches to increase the sensitivity of in situ hybridization

In our routine protocol, the sites of HRP are visualized with hydrogen peroxide and 3,3'-diaminobenzidine (DAB) in the presence of nickel and cobalt ions (Adams 1981), which increases the sensitivity more than ten-fold as compared to that of hydrogen peroxide and DAB alone. Under our conditions detailed previously (Koji and Nakane 1990), 1 - 10 pg of DNA/spot was usually detected in dot-blot hybridization. However, it was sometimes felt that the sensitivity was still not enough. We attempted various signal enhancing protocols including silver enhancement and colloidal gold-silver enhancement, but such trials were not successful in our hands. Triple layered method, where the reaction with the first antibody is followed by the second anti-

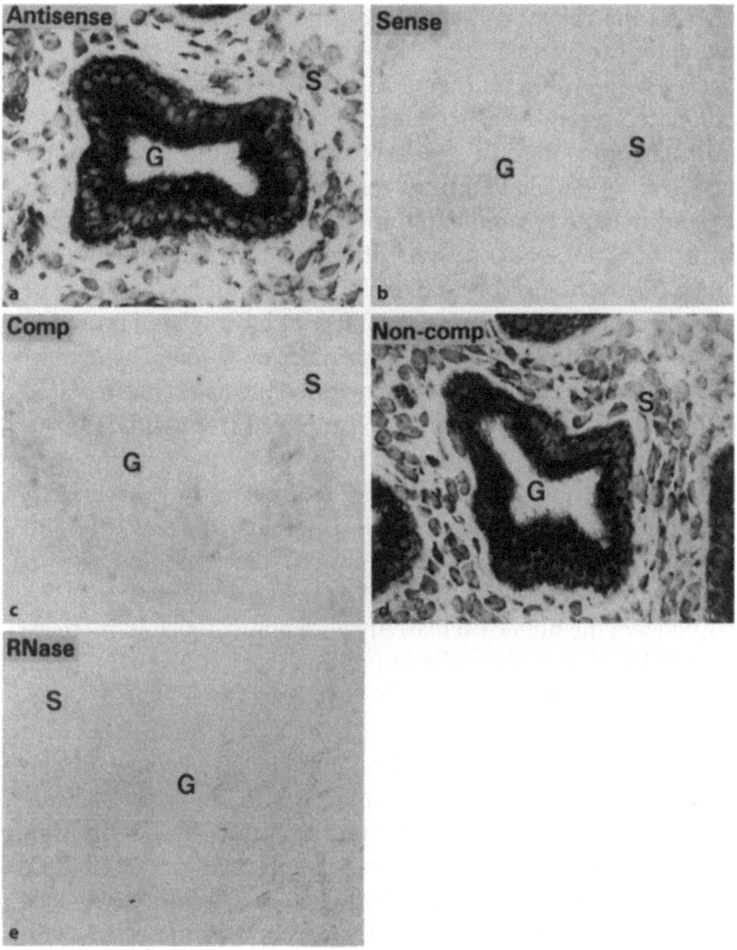

Fig. 2. An example of a set of control experiments (Koji and Brenner 1993). The set was performed to verify the specificity of the signals for estrogen receptor (ER) mRNA in the frozen sections of monkey uterus. (a) Antisense: A section was hybridized with Dig-ER antisense oligo-DNA. Strong staining for ER mRNA was found in glandular epithelial cells as well as stromal cells. (b) Sense: An adjacent section to the section hybridized with the antisense probe was hybridized with Dig-ER sense oligo- DNA. No staining was found. (c) Competition: An adjacent section was hybridized with Dig-ER antisense probe in the presence of an excess amount (33-fold) of non-labeled ER antisense oligo-DNA. The signal for ER mRNA was markedly reduced. (d) Non-competition: When an adjacent section was hybridized with Dig-ER antisense probe in the presence of an excess amount (33-fold) of non-labeled unrelated oligo-DNA. The intensity of ER mRNA signal was not changed, compared to that of "competition test". (e) RNase: A section was digested with RNase A before hybridization, and then hybridized with Dig-ER antisense probe. The section was not stained essentially. G; gland, S; stroma

body and the third HRP-labeled antibody, seemed effective to some extent (Miyazaki et al. 1996).

Very recently, Kerstens et al. (1995) have reported the use of biotinylated tyramine as a system of catalyzed reporter deposition in chromosomal in situ hybridization. More recently, De Haas et al. (1996) have found the effectiveness of the system with biotinylated tyramine in fluorescence in situ hybridization for specific RNA sequences in cultured cells. In our trial, we examined the usefulness of catalyzed signal amplification (CSA) or catalyzed reporter deposition system with biotinylated tyramine to amplify colorimetric in situ hybridization signals, using part of a DAKO kit. Our CSA protocol included the use of biotinylated tyramine after the reaction of T-T dimerized DNA with HRP-labeled anti-(T-T dimer) antibody, followed by the reaction with HRP-labeled streptavidin. When T-T dimerized λ phage DNA was spotted onto a nitrocellulose filter at 1 pg - 10 ng and then detected by the direct immunoperoxidase method, the indirect immunoperoxidase method or the CSA method, the CSA method gave the highest sensitivity with about 100-fold increase compared to that of the indirect method (Koji et al. 1997). Next, in the frozen sections of rat brain, which was fixed with 4% paraformaldehyde, 28S rRNA staining by in situ hybridization with only 10 ng/ml of T-T dimerized oligonucleotide probe complementary to rat 28S rRNA was compared with the direct method, the HRP-labeled avidin-biotin complex method and the CSA method. Again, intense 28S rRNA signal was obtained only with the CSA method, as shown in Fig. 3. Therefore, we confirmed the usefulness of the catalyzed deposition system with biotinylated tyramine in nonradioactive colorimetric in situ hybridization for specific RNA sequences in tissue sections. Further studies will be necessary to manage the development of background staining for a wide application.

When a probe labeled with some fluorescent dye is used, one can expect a direct detection of the signal by a fluorescent microscope or a laser scanning microscopy. Actually, it has turned out to be very efficient in chromosomal in situ hybridization for localization of specific sequences of genomic DNA in chromosomes (Ried et al. 1992). Very interestingly, however, a similar approach has often failed to localize specific mRNA, partly because of high background fluorescence. Therefore, even if a fluorochrome-labeled probe is used, the signal detection is

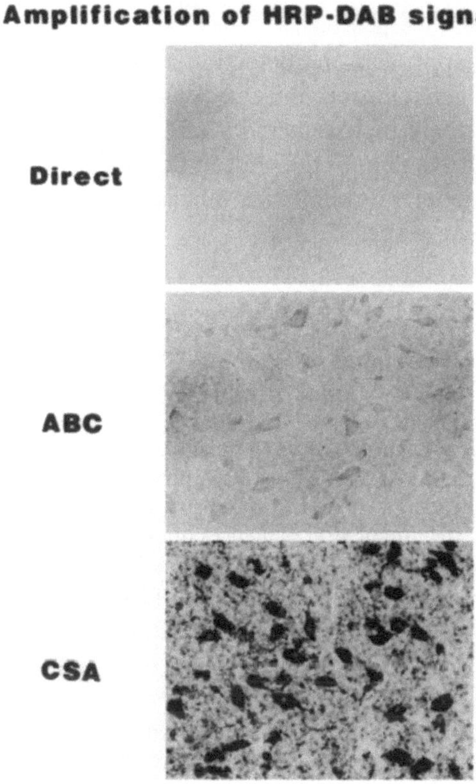

Fig. 3. Remarkable enhancement of 28S rRNA signal by catalyzed signal amplification (CSA) (Koji et al. 1997). The frozen sections of rat brain fixed with 4% paraformaldehyde in PBS were hybridized with a very low concentration of T-T dimerized oligo-DNA complementary to rat 28S rRNA. In the immunohistochemical steps, the signal was detected by the different methods. In all methods, the sites of HRP were visualized according to Adams' method (1981). Top panel (Direct): The signal was detected by the direct method with HRP-labeled anti-(T-T dimer). The staining for 28S rRNA was barely detectable. Middle panel (ABC): After hybridization, the signal was detected using the HRP- labeled avidin-biotin complex (ABC) method. The intensity of 28S rRNA signal was increased slightly but significantly. Bottom panel (CSA): After hybridization, the section was reacted with HRP-labeled anti-(T-T dimer), biotinylated tyramine and HRP-labeled streptavidin, successively. Now the signal was very clear.

usually performed enzyme-immunohistochemically with anti-fluorochrome antibody.

As a different line of approach, the detection system using chemiluminescence has been reported (Bijl et al. 1994). Although they demonstrated extremely high sensitivity and easy signal quantification, more improvement of the photon-collecting system will be needed for better morphological resolution and is maybe beyond our present scope.

References

Adams JC (1981) Heavy metal intensification of DAB-based HRP reaction products. J Histochem Cytochem 29:775

Bijl JJ, Rieger E, van Oostveen JW, Meijer CJLM, Oudejans CBM, Walboomers JMM (1994) Quantification of biotinylated RNA probes for in situ hybridization using chemiluminescence. Histochemistry 102:77-82

Brahic M, Haase AT (1978) Detection of viral sequences of low reiteration frequency by in situ hybridization. Proc Natl Acad Sci USA 75:6125-6129

De Haas RR, Verwoerd NP, van der Corput MP, van Gijlswijk RP, Siitari H, Tanke HJ (1996) The use of peroxidase-mediated deposition of biotin-tyramide in combination with time-resolved fluorescence imaging of europium chelate label in immunohistochemistry and in situ hybridization. J Histochem Cytochem 44:1091-1099

Ehara H, Deguchi T, Koji T, Yang M, Ito S-I, Kawada Y (1996) Autoclave antigen retrieval technique for immunohistochemical staining of androgen receptor in formalin-fixed paraffin sections of human prostate. Acta Histochem Cytochem 29:311-318

Gall JG, Pardue M (1969) Formation and detection of RNA-DNA hybrid molecules in cytological preparations. Proc Natl Acad Sci USA 63:378-383

Goto M, Koji T, Mizuno K, Tamaru M, Koikeda S, Nakane PK, Mori N, Masamune Y, Nakanishi Y (1990) Transcription switch of two phosphoglycerate kinase genes during spermatogenesis as determined with mouse testis sections in situ. Exp Cell Res 186:273-278

Hakuno N, Koji T, Yano T, Kobayashi N, Tsusumi O, Taketani Y, Nakane PK (1996) Fas/APO-1/CD95 system as a mediator of granulosa cell apoptosis in ovarian follicle atresia. Endocrinology 137:1938-1948

Henderson GS, Conary JT, Davidson JM, Stewart SJ, House FS, McCurley TL (1991) A reliable method for northern blot analysis using synthetic oligonucleotide probes. BioTechniques 10:190-197

John HA, Birnstiel ML, Jones KW (1969) RNA-DNA hybrids at the cytological level. Nature 223:582-587

Koji T, Brenner RM (1993) Localization of estrogen receptor messenger ribonucleic acid in rhesus monkey uterus by nonradioactive in situ hybridization with digoxigenin-labeled oligodeoxynucleotides. Endocrinology 132:382-392

Koji T, Nakane PK (1990) Localization in situ of specific mRNA using thymine-thymine dimerized DNA probes. Sensitive and reliable non-radioactive in situ hybridization. Acta Pathol Jpn 40:793-807
Koji T, Nakane PK (1996) Recent advances in molecular histochemical techniques: in situ hybridization and southwestern histochemistry. J Electron Microsc 45:119-127
Koji T, Moriuchi T, Nakane PK (1988) Improved tissue preparation for in situ localization of specific mRNA using non-radioactive DNA probes: Effects of protease digestion and probe size on signal detection in frozen and paraffin sections of rat pituitary glands. Acta Histochem Cytochem 21:187-200
Koji T, Sugawara I, Kimura M, Nakane PK (1989) In situ localization of c-myc mRNA in HL-60 cells using non-radioactive synthetic oligodeoxynucleotide probes. Acta Histochem Cytochem 22:295-307
Koji T, Chedid M, Rubin JS, Slayden OD, Csaky KG, Aaronson SA, Brenner RM (1994) Progesterone-dependent expression of keratinocyte growth factor mRNA in stromal cells of the primate endometrium: Keratinocyte growth factor as a progestomedin. J Cell Biol 125:393-401
Koji T, Kanemitsu Y, Hoshino A, Nakane PK (1997) A novel amplification method of nonradioactive in situ hybridization signal for specific RNA with biotinylated tyramine. Acta Histochem Cytochem 30:401-406
Lawernce JB, Singer RH (1985) Quantitative analysis of in situ hybridization methods for the detection of actin gene expression. Nucl Acids Res 13:1777-1799
Meinkoth J, Wahl G (1984) Hybridization of nucleic acids immobilized on solid supports. Anal Biochem 138:267-284
Miyazaki M, Abe K, Koji T, Furusu A, Ozono Y, Harada T, Nakane PK, Yagame M, Endoh M, Nomoto Y, Sakai H (1996) Intraglomerular C3 synthesis in human kidney detected by in situ hybridization. J Am Soc Nephrol 7:2428-2433
Moench TR, Gendelman HE, Clements JE, Narayan O, Griffin DE (1985) Efficiency of in situ hybridization as a function of probe size and fixation technique. J Virol Method 11:119-130
Nakane PK, Matsumura H, Koji T (1990) In situ hybridization using T-T dimerized non-radioactive probes. In: Tabuchi K, ed. Proceedings of the 8th Nikko Brain Tumor Conference. Tokyo: Springer-Verlag, 52-62
Razzaque MS, Koji T, Taguchi T, Harada T, Nakane PK (1994) In situ localization of type III and type IV collagen-expressing cells in human diabetic nephropathy. J Pathol 174:131-138
Ried T, Baldini A, Rand TC, Ward DC (1992) Simultaneous visualization of seven different DNA probes by in situ hybridization using combinatorial fluorescence and digital imaging microscopy. Proc Natl Acad Sci USA 89: 1388-1392
Ruizeveld de Winter JA, Trapman J, Vermey M, Mulder E, Zegers ND, van der Kwast, TH (1991) Androgen receptor expression in human tissues: An immunohistochemical study. J Histochem Cytochem 39:927-936
Slayden OD, Koji T, Brenner RM (1995) Microwave-stabilization enhances immunocytochemical detection of estrogen receptor in frozen sections of macaque oviduct. Endocrinology 136: 4012-4021

Thomas CA, Dancis BM (1973) Ring stability. J Mol Biol 77:44-55
Ueki T, Koji T, Tamiya S, Nakane PK, Tsuneyoshi M (1995) Expression of both basic fibroblast growth factor and fibroblast growth factor receptor indicates a poor prognosis in human gastric carcinoma. J Pathol 177:353-361
Wetmur JG, Davidson N (1968) Kinetics of renaturation of DNA. J Mol Biol 31:349-370
Yoshii A, Koji T, Ohsawa N, Nakane PK (1995) In situ localization of ribosomal RNAs is a reliable reference for hybridizable RNA in tissue sections. J Histochem Cytochem 43:321-327

Chapter 6

In Situ Hybridization for DNA: Fluorescent Probe

MORIMASA MATSUTA AND MAYUMI MATSUTA

Introduction

Fluorescence in situ hybridization (FISH) with chromosome-specific probes is one of the non-isotopic in situ hybridization methods that opens up several new areas of cytogenetic investigation by allowing visual determination of the presence and normality of specific genetic sequences in single metaphase or interphase cells (Pinkel 1986a, 1986b). With this approach, the genetic loci to be analyzed are made microscopically visible in single cells using in situ hybridization with nucleic acid probes specific to these loci. The DNA in target cells is made as a single strand and incubated with single-stranded fluorochrome-directly-labeled or chemically modified DNA probe under conditions where the probe will anneal only with DNA sequences to which it has high DNA sequence homology. The bound chemically modified probe is made visible by immunofluorescent detection methods. This technique is often called standard FISH.

Multi-target FISH (M-FISH) provides the means of detecting multiple genetic targets simultaneously within individual nuclei or metaphase spreads. By combining probes for a number of targets into a single hybridization, precious specimens are conserved, assay time and labor are reduced, and the clonal or mosaic nature of a specimen can be accurately assayed (VYSIONS 1997).

A recently described M-FISH technology, 24-color FISH using spectral karyotyping (SKY), permits the simultaneous visualiza-

Correspondence to Morimasa Matsuta, Iwate Medical University, Department of Obstetrics and Gynecology, 19-1 Uchimaru, Morioka, 020-8505, Japan (*phone* +81-19-651-5111 extension 2344; *fax* +81-19-622-1900; *e-mail* matsutao@iwate-med.ac.jp)

tion of all human chromosomes in 24 different colors. This method demostrates the characterization of de novo structural chromosomal abnormalities (Schroeck et al. 1997).

The newly developed cytogenetic technique, comparative genomic hybridization (CGH), produces a map of the DNA sequence copy number as a function of chromosomal location throughout the entire genome (Kallioniemi 1994,1993,1992). Test DNA and normal reference DNA are differentially labeled with two fluorochromes, and the DNAs are hybridized simultaneously to normal chromosome spreads. Regions of gain or loss of DNA, such as deletions, duplications, or amplifications, are seen as changes in the ratio of the intensities of the two fluorochromes along the target chromosomes.

Preimplantation, prenatal, and postnatal genetic diagnosis

Preimplantation and prenatal diagnosis is a procedure for the diagnosis of genetic diseases by screening pregnancies or in vitro-fertilized embryos for chromosome abnormalities. A number of techniques are used for the specific diagnosis of inherited disorders, such as the polymerase chain reaction. FISH can be used for the preimplantation, prenatal, and postnatal diagnosis of abnormal numbers of chromosomes and chromosome rearrangement.

Preimplantation genetic diagnosis is performed when the embryo is at the 6-10 cell stage when embryos are generated by standard in vitro fertilization. One or two blastmeres are obtained from the cleavage-stage embryo biopsy, and are used for diagnosis.

Prenatal genetic diagnosis is performed by chorionic villi sampling (CVS) or amniocentesis. Method for the application of FISH to cells derived from CVS and is performed at 10-12 weeks of pregnancy. Controlled studies are necessary, however, to appreciate the impact of the moderately common condition of chromosomal mosaicism found in CVS samples. Amniocentesis involves the aspiration of a small volume of amniotic fluid. Chromosomes are analyzed by karyotyping, although it can take more than 3 weeks to obtain the relevant genetic information. FISH allows for the rapid detection of major genetic abnormalities, which will eliminate the anxious waiting weeks. This technique can also be applied to fetal blood cells by cord-centesis, and in the near future it will be applied to fetal cells in maternal blood.

Postnatal genetic diagnosis is performed by the application of FISH to the nuclei of peripheral blood cells.

FISH using the chromosome-specific probes of chromosomes 13, 18, 21, X, and Y detects the majority of trisomies and monosomies.

Microdeletions, such as Prader-Willi/Angelman, DiGeorge/VCFS regions, are small deletions of genes or regions that are difficult to identify by karyotyping. Using probes for these genes or regions, FISH can detect whether or not these microdeletions are present.

Development of a CGH application to a single cell is useful for cytogenetic analysis.

Oncology

Many genetic changes underlie the development and progression of tumors.

- Chromosomal numerical aberrations are frequently observed during the development of tumors, and some of them contribute to tumor progression while others cause genetic instability in tumor cells. Chromosomal numerical aberrations may be easily detected in precancerous and malignant lesions by FISH using chromosome-specific probes in interphase cells or tissue sections.
- FISH can identify the morphologically normal leukemic cells after anti-leukemic therapy, since leukemic cells show signs of aneusomy. The technique may be applied to determine whether bone marrow transplantation was successful, by identifying the blood cells in a female patient originated from male donor.
- Translocations, such as Philadelphia chromosome, t(12;21), etc., can be detected in interphase nuclei by dual-color FISH using the probes for one or more whole chromosomes or chromosomal segments.
- Deletion of tumor suppressor genes, such as p53 and Rb etc., or amplification of oncogenes, such as c-myc and HER-2/neu

etc., can be detected in interphase nuclei by dual-color FISH using a combination of chromosome-specific and gene-specific probes.

- CGH provides an overview of DNA sequence copy number changes (losses, deletions, gains, amplifications) in a tumor specimen and maps these changes on normal chromosomes, in a single hybridization. CGH is in particular useful for the analysis of DNA sequence copy number changes in common solid tumors where complex karyotypes with numerous markers, double minutes, and homogeneously stained chromosomal regions are common. In other words, this technique is useful for screening both unknown and known oncogenes. Application of CGH in diagnosis and prognostic evaluation of tumors is promising.

Others

The quantification and/or cell cycle analysis of FISH-samples via flow cytometry has been reported. Three-dimensional organization of chromosomes in interphase nuclei, which may change with the cell cycle, can be analyzed using a confocal laser scanning microscope and image analysis devices.

Subprotocol 1
Fluorescence In Situ Hybridization (FISH)

These protocols have demonstrated the ability to detect hybridization of probes with target size of approx. 3 kb to whole genomes.

Outline

A short protocol of the entire FISH procedure is given here:

1. Denature specimen slides in 70% formamide/2X SSC at 74°C for 5 mins.

2. Dehydrate specimen slides in ethanol series in order to maintain denatured DNA single strands in specimen.
3. Make hybridization mix and denature the hybridization mix at 74°C for 5 mins.
4. Apply the denatured hybridization mix to specimen target area on slides.
5. Hybridize probe DNA to target specimen DNA at 37°C for 30 mins to 2 days.
6. Wash to remove non-specifically bound probe in 50% formamide/2X SSC at 46°C, followed by buffer wash. (If the probes are modified with haptens, such as biotin or digoxigenin etc., this step should be followed by fluorescent staining procedures.)
7. Air dry slides, and apply counterstain.
8. View using fluorescence microscope with a suitable filter set.

Slides containing cells and chromosomes are prepared using standard technique with methanol-acetic acid (3:1) fixation. Paraformaldehyde (1-4% in phosphate buffer, pH 7.4, for 10-30 mins at 4°C) or ethanol fixation also yields good signals. However, the signals are much better when the cells do not have cytoplasm. After fixation, slides can be stored at -20°C for up to 6 months.

Formalin-fixed paraffin embedded tissue sections are also used for FISH assay. After deparaffinization and rehydration, the sections are treated as described in standard FISH protocol using pepsin or other proteolytic enzymes at high concentration before or after denaturation. The digestive conditions vary among specimen types, and need to be optimized in each case.

Materials

- Fluorescent microscope (e.g., Olympus BX type, Nikon) with suitable single-bandpass or multi-bandpass filter sets (for FITC, Rhodamin, Texas-Red, Spectrum-Orange, Spectrum-Green, DAPI, PI, etc.).
- Water Baths: 37°C, 46°C and 74°C
- Vortex

- A bench-top microcentrifuge
- Microcentrifuge tubes: 0.5 ml, 1 ml
- Diamond tipped pen
- Glass slides: aminoalkylsilane-coated, commercially available.
- Glass cover slips; 18X18 mm, 22X40~60 mm
- Plastic cover slips; 22X40~60 mm, cut from polyethylene bag
- Coplin jars
- Micropipettes (variable type:0.5~10μl, ~200μl, 1ml)
- Moist chamber
- Incubator at 37°C
- Hot plate for Rapid Hybridization and Codenaturation procedures
- Chromosome-specific, locus-specific probes; also commercially available from Vysis (Downers Grove, IL), Oncor (Geithersburg, MD), etc.
- Master Mix (of Hybridization): Master Mix gives a hybridization mixture that is 55% formamide/10% dextran sulfate/ 1XSSC.
- Purified water
- Formamide: Ultra pure grade (Kodak etc.)
- Rubber cement: Paper bond™ (Kokuyo, Tokyo, Japan)
- Denaturation solution: 50 ml of 70% formamide/2X SSC
- Washing solution: 150 ml of 50% fromamide/2X SSC
- Washing solution for rapid hybridization protocol: 0.4X SSC
- 20X SSC solution
- PN buffer
- PNM buffer: 5% nonfat dry milk to PN buffer + 0.02% NaAzide.
- Antifade solution (see Recipes, also commercially available)
- 2'-6-diamino-2-phenyl-indole (DAPI): 100 ng (- 1000 ng for tissue sections) in antifade solution; this is required when either green-fluorochrome (FITC, Spectrum Green, etc.)-labelled probes or red-fluorochrome (TexasRed, Sprectrum Orange, etc.)- labelled probes are used as counterstains
- Propidium iodide (PI): 5 - 10 μg in antifade solution, counterstain for green-fluorochrome (FITC, Spectrum Green, etc.) - labelled probes.
- Ethanol dehydrate solution: concentrations of 70, 85 and 100%.

Fluorescent staining reagents

- Fluorescein-avidin (Vector Laboratories, A-2011)
- Biotinylated Anti-avidin (Vector Laboratories, BA-0300)
- Texas-Red-avidin (Vector Laboratories, A-2016)
- Anti-digoxigen-fluorescein, Fab fragments (Boehringer Mannheim, 1207 741)
- Anti-digoxigen-rhodamine, Fab fragments (Boehringer Mannheim, 1207 750)

Recipes

- Aminoalkylsilane (APES) coated glass slides:
  - 1 ml of 3-aminopropylriethoxysilane in 50 ml acetone.
  - Slides should be dipped into this solution for 20 secs, then washed twice in acetone and twice in distilled water. Dry slides at 37°C overnight.
- Denaturation solution: 50 ml of 70% formamide/2X SSC
  - 35 ml formamide
  - 5ml 20X SSC
  - 7 ml purified water
  - Adjust pH to 7.0 with 1N HCl, adjust final volume to 50 ml with purified water. This solution can be reused for a few weeks as long as stored at 4°C.
- Washing solution: 150 ml of 50% fromamide/2X SSC
  - 75ml formamide
  - 15 ml 20X SSC
  - 60 ml purified water
  - Adjust pH to 7.0 with 1N HCl. This solution can be reused for a few weeks as long as stored at 4°C.
- Washing solution for sub-protocol 2: 0.4X SSC
  - Add 20ml 20X SSC to 950 ml purified water. Adjust pH to 7.0 with 1N NaOH. Add purified water to bring final volume of the solution to 1 liter. Store up to six months at room temperature. The washing solution must be at 74°C before using. In order to maintain the proper temperature, four slides must be placed in the heated solution at one time. If fewer than four slides have been hybridized, room temperature microscope slides must be used to bring the number to four slides. If more than four slides have been hybridized they must be washed in more than one Coplin jar. The temperature of the wash solution must return to 74°C before washing each jar. Do not remove the coverslips from several slides before placing any of the slides in the wash jar. Begin timing the incubation when the last slide has been added to the wash bath.

- 20X SSC solution
  - 175.3 g NaCl (MW 58.44)
  - 88.2 g Na citrate (MW 294.1)
  - Volume to 800 ml with distilled water.
  - Adjust pH to 7.0 with 1M HCl.
  - Adjust final volume to 1 liter.
  - Autoclave if desired.
- PN buffer: 10 liter
  - Make 10 liters 0.1 M Na_2HPO_4 (dibasic). If using heptahydrate, MW 268.07: 268.07 g/10 liters distilled water.
  - Make 600 ml 0.1 M Na_2HPO_4 (monobasic). If using monohydrate, MW 137.99: 8.28 g/600ml distilled water.
  - Titrate the dibasic solution (pH>9) by adding the monobasic solution (pH ~4.5) until a pH of 8.0 is reached. Keep track of the volume of monobasic solution added; not all of it will be used.
  - Add Nonidet P-40 (NP-40) to 0.05% (should be approximately 5.25 ml). NP-40 does not influence the pH reading. This buffer can be stored at room temperature indefinitely.
- PNM buffer:
  - Add 5% nonfat dry milk to PN buffer + 0.02% NaAzide, heat to 37°C, and leave on lab bench overnight, perhaps longer. Spin in a tabletop centrifuge to pellet the solids. Remove the supernant and store at 4°C or -20°C. If more solids form, repeat spin.
- Antifade:
  - 100 mg p-phenylenediamine dihydrochloride (Sigma p1519) in 10 ml PBS
  - Adjust to pH 8 with 0.5 M Carbonate-bicarbonate buffer (0.42 g $NaHCO_3$ in 10 ml water, pH to 9 with NaOH).
  - Add to 90 ml glycerol. Filter through 0.22 μm to remove undissolved particulates. Store in aliquot with the desired counterstains in the dark at -20°C. This solution darkens with time, but still remains effective. It can be used after more than 6 months storage. (J. Immuno. Method 43, 349, 1981)
- Ethanol dehydrate solution
  - For final concentrations of 70, 85 and 100%. Prepare v/v dilutions of 100% ethanol with purified water. Use solution for up to seven days and then discard. If solution evaporates or becomes diluted, replace with fresh solution. Between periods of use, store at room temperature.

- Master Mix (of Hybridization)
  - 5.5 ml formamide
  - 1g dextran sulfate
  - 0.5 ml 20X SSC
  - These are heated to 70°C for several hours to dissolve the dextran sulfate, cooled, pH adjusted to 7.0 and the volume brought to 7 ml. This volume is 70% of that used in the final hybridization mix. The remaining 30% of the volume are filled with probe(s), carrier DNA, and water if needed. Used in this way Master Mix gives a hybridization mixture that is 55% formamide/10% dextran sulfate/1XSSC.

Procedure

1. Bring Coplin jar containing the denaturation solution (70% formamide/2XSSC) to room temperature. Place the jar in a 74°C water bath 30 mins prior to use in order to attain the solution temperature. Measure the temperatures of the solution inside the Coplin jar, and make sure the temperature is ≤72°C.
2. Thaw reagents at room temperature, then centrifuge each tube for a few secs using a bench-top microcentrifuge.
3. Prepare ethanol series in three Coplin jars: 50ml 100% ethanol into one jar, 50ml 85% ethanol into another, and 50ml 70% ethanol into the last.

FISH using fluorochrome-directly-labeled DNA probes

Note: Perform all steps, which do not require light for manipulation (incubation periods, washes, etc.), in darkness in order to avoid photo bleaching of fluorochrome-directly-labeled probes.

Remarks

Slides containing cells are prepared using methanol-acetic acid (3:1) fixation.

Heat or enzymatic treatment may induce loss of adherence of cells and tissue sections to glass slides. This may be overcome by using aminoalkylsilane-coated glass slides.

For single copy hybridization, the slides are treated with RNAse, Proteinase K, and paraformaldehyde. Apply 200 μl of RNAse (100 μg/ml in 2XSSC) to each slide and cover with a large plastic coverslip (24X50 ~ 60 mm). Place slides in a moist chamber at 37°C for one hour. Rinse in four changes of 2XSSC at room temperature, dehydrate in an ethanol series and air dry. The slides are then treated with Proteinase K for 7.5 mins at 37°C, washed in 2XSSC, fixed in 4% paraformaldehyde in phosphate buffer (PB) for 10 mins, and washed in 2XSSC. The Proteinase K concentration is high enough so that without the fixation the chromosomes would be destroyed in denaturation procedure, suggesting concentration at 0.6 - 6μg/ml in 20 mM Tris/2 mM $CaCl_2$ pH 7.5.

Slide preparation

Note: Immerse the slides in 2XSSC for 30 mins at 37°C in order to "age" them, if the specimen were freshly prepared within the last 24 hours. Dehydrate the slides through a series of 1 minute ethanol washes (70%, 85%, 100%). Air dry. This may help chromosomes maintain morphology during the hybridization.

1. Locate best area on slide using low power microscopy. Mark the back of the slide with a diamond scribe to indicate roughly where the boundaries of the coverslip should be. A typical hybridization area might be a square of 20 mm. The hybridization and multiple cytochemical reactions are carried out under coverslips to minimize the amount of reagents required.

2. Immerse the slides in the 74°C denaturation solution (70% formamide/2XSSC) for 5 mins.

Note: A room temperature slide put into 50 ml Coplin jar will cause the temperature to drop by about 1°C, so do only a few slides at a time.

3. Transfer the slide quickly to 70% ethanol for 1 minute. Agitate to rinse off denaturing solution.

4. Continue dehydration in 85% ethanol for 1 minute, and 1 minute in 100% ethanol.

5. Remove the slides from the 100% ethanol.

6. Dry slides by touching the bottom edge of the slides to a blotter and wiping the underside of the slides dry with a paper towel.

7. Air-dry.

Hybridization mix

8. Add the following to a 0.5 ml microcentrifuge tube at room temperature:
   - 7μl of Master Mix
   - 1μl carrier DNA (herring testis DNA or Cot-l DNA:500μg/ml)
   - 2μl probe (fluorochrome-directly-labeled DNA probe: diluted if necessary)total 10μl , for each target area.

Note: Master Mix (of Hybridization) volume is 70% of that used in the final hybridization mix. The remaining 30% of the volume are filled with probe(s), carrier DNA, and water if needed. Used in this way Master Mix gives a hybridization mix that is 55% formamide/10% dextran sulfate/1XSSC. To simultaneously hybridize up to three probes, each labeled with different fluorochromes, add 1 μl of each probe. Add purified water to bring the combined volume of probe and water to 3 μl.
Hybridization buffer, which contains both Master Mixture and carrier DNA, is commercially available from Vysis or Oncor. Typical probe concentrations are 2 μg/ml for genomic DNA and 0.4μg/ml for chromosome-specific repetitive sequences. Use 2 - 3 μl of hybridization mix /cm^3. Thus a typical hybridization mix for a 18-20mm square area on a slide might be 10 μl.

9. Centrifuge tube for a few secs.

10. Vortex and re-centrifuge.

11. Place tube in a 74°C water bath for 5 mins.

12. Remove tube from water bath.

13. Place tube on ice until ready to apply hybridization mix to target DNA on slides.

Note: If slides are ready, you can apply hybridization mix immediately to target area.

Hybridization

14. Apply 10 μl of hybridization mix to slide. Apply 18 - 22 mm square glass coverslip on placing hybridization mix.

15. Seal coverslips with rubber cement.

16. Place slides in a prewarmed humidified box and place box in a 37°C incubator for 6-16 hours.

Note: To produce an assay with sufficient signal, start with over night hybridization for most probes. With repetitive probes the hybridization incubation can be shortened to 30 mins (signals are visible after 10 mins). The hybridization signal reaches a maximum after about 6-8 hours.

Washing the slide

Prepare 3 Coplin jars of 50% formamide/2XSSC wash solution for washing, and a jar of 2X SSC. Place the jar in a 46°C water bath 30 mins prior to use in order to attain the solution temperature.

17. Remove slides from incubator and peel off rubber cement with forceps, and immediately immerse the slides in jar #1 of 50% formamide/2XSSC wash solution at 46°C. Agitate slides for 1-3 secs. The coverslips can be slid off easily in the first wash. Do this gently so the specimens are not scratched. Remove slides after 10 mins.
18. Immerse slides in jar #2 of wash solution at 46°C. Agitate slides for 1-3 secs. Remove slides after 10 mins.
19. Immerse slides in jar #3 of wash solution at 46°C. Agitate slides for 1-3 secs. Remove slides after 10 mins.
20. Immerse slides in 2XSSC at 46°C. Agitate slides for 1-3 secs. Remove slides after 10 mins.
21. Immerse slides in 2XSSC/0.1% Nonidet P-40 (NP-40) at room temperature for 1 minute.
22. Allow slides to air dry in darkness.
23. Mount in an anti-fade solution containing DAPI or propidium iodide (PI) for green fluorochromes (FITC, Spectrum-Green, etc.). View using fluorescent microscope.

Note: You may apply the rapid wash procedure in Protocol for Rapid Procedure.

FISH using chemically modified DNA probes

Note: If the probes are modified with haptens, such as biotin or digoxigenin etc., fluorescent staining procedures should follow. These probes yield brighter signals than fluorochrome-directly-labeled probes with slightly high background.

Remarks

Slides containing cells are prepared using methanol-acetic acid (3:1) fixation.

Heat or enzymatic treatment may induce loss of adherence of cells and tissue sections to glass slides. This may be overcome by using aminoalkylsilane-coated glass slides.

For single copy hybridization, the slides are treated with RNAse, Proteinase K, and paraformaldehyde. Apply 200 µl of RNAse (100 µg/ml in 2XSSC) to each slide and cover with a large plastic coverslip (24 X 50 ~ 60 mm). Place slides in a moist chamber at 37°C for one hour. Rinse in four changes of 2XSSC at room temperature, dehydrate in an ethanol series and air dry. The slides are then treated with Proteinase K for 7.5 mins at 37°C, washed in 2XSSC, fixed in 4% paraformaldehyde /PBS for 10 mins, and washed in 2XSSC. The Proteinase K concentration is high enough so that without the fixation the chromosomes would be destroyed in denaturation procedure, suggesting concentration at 0.6 - 6µg/ml in 20 mM Tris/2 mM $CaCl_2$ pH 7.5.

Slide preparation for cell specimen

1. Immerse the slides in 2 x SSC for 30mins at 37°C in order to "age" them, if specimens were freshly prepared within the last 24 hrs. Dehydrate the slides through a series of diluted ethanol. Air dry.

2. Locate best area on slide using low power microscopy. Mark the back of the slide with a diamond scribe to indicate roughly where the boundaries of the coverslip should be. A typical hybridizaiton area might be 20 mm square. The hybridization and multiple cytochemical reactions are carried out under coverslips to minimize the amount of reagents required.

3. Immerse the slides in the 74°C denaturation solution (70% formamide/2XSSC) for 5 mins.

Note: A room temperature slide put into 50 ml Coplin jar will cause the temperature to drop by about 1°C, so do only a few slides at a time.

4. Transfer the slide quickly to 70% ethanol for 1 minute. Agitate to rinse off denaturing solution.

5. Continue dehydration in 85% ethanol for 1 minute, and 1 minute in 100% ethanol.

6. Remove the slides from the 100% ethanol.

7. Dry slides by touching the bottom edge of the slides to a blotter and wiping the underside of the slides dry with a paper towel.

8. Air-dry.

Hybridization mix

9. Add the following to a 0.5 ml microcentrifuge tube at room temperature:
   - 7μl of Master Mix
   - 1μl carrier DNA (herring testis DNA or Cot-l DNA:500μg/ml)
   - 2μl probe (biotinized or digoxigenin-labeled DNA probe: diluted if necessary)total 10μl , for each target area

Note: Master Mix (of Hybridization) volume is 70% of that used in the final hybridization mix. The remaining 30% of the volume are filled with probe(s), carrier DNA, and water if needed. Used in this way Master Mix gives a hybridization mix that is 55% formamide/10% dextran sulfate/1XSSC. Hybridization buffer, which contains both Master Mix and carrier DNA, is commercially available from Vysis or Oncor.
Typical probe concentrations are 2 μg/ml for genomic DNA and 0.4μg/ml for chromosome-specific repetitive sequences. A typical hybridization mix for a 18-20mm square area on a slide might be 10 μl.

10. Centrifuge tube for a few secs.

11. Vortex and re-centrifuge.

12. Place tube in a 74°C water bath for 5 mins.

13. Remove tube from water bath.

14. Place tube on ice until ready to apply hybridization mix to target DNA on slides.

Note: If slides are ready, you can apply hybridization mix immediately to target area.

Hybridization

15. Apply 10 μl of probe mix to slide. Apply 18-22 mm square glass coverslip on placing hybridization mix.

16. Seal coverslips with rubber cement.

17. Place slides in a prewarmed humidified box and place box in a 37°C incubator for 6-16 hours.

Note: To produce an assay with sufficient signal, start with over night hybridization for most probes. With repetitive probes the hybridization incubation can be shortened to 30 mins (signals are visible after 10 mins). The hybridization signal reaches a maximum after about 6-8 hours.

Washing the slide

Prepare 3 Coplin jars of 50% formamide/2XSSC wash solution for washing, and a jar of 2X SSC. Place the jar in a 46°C water bath 30 mins prior to use in order to attain the solution temperature.

18. Remove slides from incubator and peel off rubber cement with forceps, and immediately immerse the slides in jar #1 of 50% formamide/2XSSC wash solution at 46°C. Agitate slides for 1-3 secs. The coverslips can be slid off easily in the first wash. Do this gently so the specimens are not scratched. Remove slides after 10 mins.

19. Immerse slides in jar #2 of wash solution at 46°C. Agitate slides for 1-3 secs. Remove slides after 10 mins.

20. Immerse slides in jar #3 of wash solution at 46°C. Agitate slides for 1-3 secs. Remove slides after 10 mins.

21. Immerse slides in 2XSSC at 46°C. Agitate slides for 1-3 secs. Remove slides after 10 mins.

22. Immerse slides in PN buffer at room temperature for 1 minute.

Note: You may apply the rapid wash procedure in Protocol for Rapid Procedure (Subprotocol 2).

Fluorescent staining using biotinized probes

Note: The slide should not be allowed to dry because the resulting salt crystals cause nonspecific binding of the detection reagents.

23. Remove a slide from PN buffer, drain and blot excess liquid form edge. Apply diluted avidin-FITC or -Texas-Red (5 µl/ml in PNM buffer, 5µl/cm^2), and cover with a plastic coverslip cut from a polyethylene bag.

24. The slide is then incubated for 20 mins in a moist chamber at room temperature in darkness.

Note: The amount of avidin used here is in excess so dilution as it is applied to the wet slide is not critical. The concentration is sufficient even for the amplification steps, where there are many times more binding sites available.

25. Wash the slides in 3 changes of PN buffer at room temperature for 2 mins each.

26. Mount in an anti-fade solution (1.5 μl/cm^2) containing DAP, or PI for FITC, and a no. 1 thickness coverslip used. View using fluorescent microscope.

Amplification of the biotinized probe fluorescence

Note: Amplification of the biotinized probe fluorescence is accomplished by applying a biotinylated goat-anti-avidin antibody followed by another layer of avidin after Step 25. This can be done even after observation of the slide after Step 26, followed by removing cover slips and washes as in Step 25 in PN buffer.

1. Drain and blot the excess liquid from the slide.
2. Add the biotinylated goat-anti-avidin antibody solution (100 times diluted in PNM buffer), and incubate at room temperature for 20 mins under a plastic coverslip.
3. Wash the slides in 3 changes of PN buffer at room temperature for 2 mins each.
4. Remove a slide from PN buffer, drain and blot excess liquid form edge.
5. Apply diluted avidin-FITC or -Texas-Red (5μ l/ml in PNM buffer, 5μl/cm^2), and cover with a plastic coverslip.
6. The slide is then incubated for 20 mins in a moist chamber at room temperature in darkness.
7. Wash the slides in 3 changes of PN buffer at room temperature for 2 mins each.
8. Mount in an anti-fade solution (1.5 μl/cm^2) containing DAPI or propidium iodide (PI) for FITC, and a no. 1 thickness coverslip used. View using fluorescent microscope.

Fluorescent staining using digoxigenin labelled probes

Follow step 22.

23. Remove a slide from PN buffer, drain and blot excess liquid form edge.
24. Apply diluted anti-digoxigenin-FITC or -TexasRed (10-20 times diluted in PNM buffer, 5μl/cm^2), and cover with a plastic coverslip cut from a polyethylene bag.
25. The slide is then incubated for 20 mins in a moist chamber at room temperature in darkness.
26. Wash the slides in 3 changes of PN buffer at room temperature for 2 mins each.
27. Mount in an anti-fade solution (1.5 μl/cm^2) containing DAP, or PI for FITC, and a no. 1 thickness coverslip used. View using fluorescent microscope.

Results

Tandem repeats in the centromeric regions of all chromosomes are specific for each chromosome. The probes for these sequences identify each chromosome and detect numerical aberrations in the nuclei. Hybridization signals are compact in interphase nuclei. Furthermore, the target sequences are very bright, since there are frequently megabases of target for short probe sequences (Fig. 1, 2). The second type of probe makes one or more whole chromosomes or chromosomal segments visible, and can be used to analyze structural abnormalities in metaphase spreads (Fig.3). The third type of probe is specific for genome loci and can be used to detect genetic events such as deletions, amplifications, inversions, and translocations in interphase nuclei (Fig. 4,5).

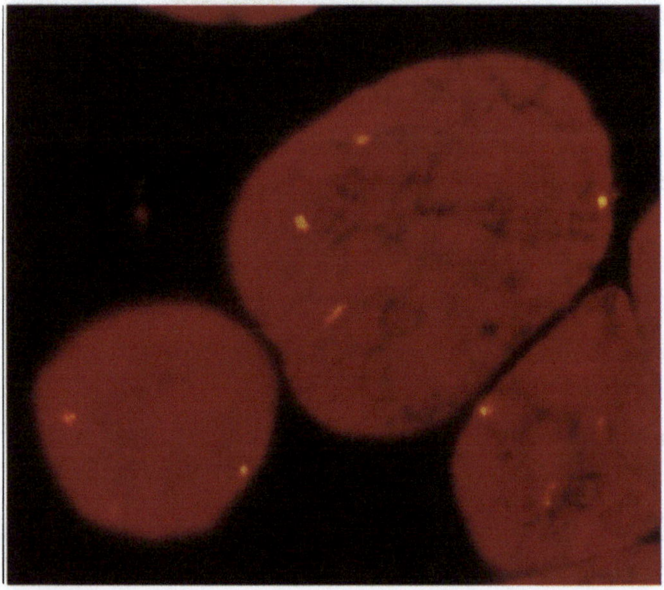

Fig. 1. FISH using a centromeric probe specific for chromosome 17. Left nucleus has 2 copies, center one has 4 copies, and right one has 3 copies. Suggesting intra-tumoral heterogeneity of tumor cells. Ovarian cancer. Touch preparation. Counterstain with propidium iodide. X 1,000

Chromosome enumeration

Counting the centromeric signals: Counting was performed using criteria previously published with slight modifications: i.e.,

- nuclei should not be overlapped,
- signal intensity should be more or less of the same homogeneous staining intensity,
- minor hybridization spots, which can be recognized by their smaller size and lower intensity, should be excluded
- signals may only be counted when completely separated from each other and paired (split phenomenon in G_2 phase) or closely opposed spots count as one signal.

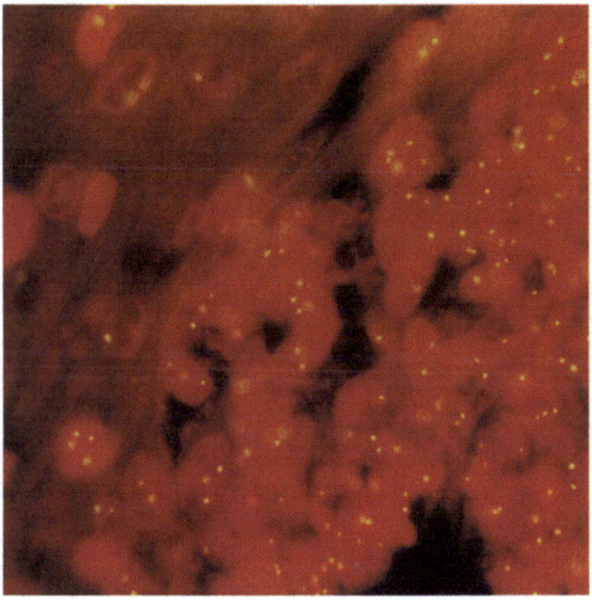

Fig. 2. FISH using a centromeric probe specific for chromosome 17. Normal keratinocytes at upper half show 0-2 signals. On the contrary, melanoma cells at lower half have several signals, showing a numerical aberration in melanoma. Formalin-fixed paraffin embedded section. Counterstain with propidium iodide. X 1,000

Gene deletions and amplification

Deletions can be detected by dual-color FISH with a combination of centromeric and locus-specific probes. Cells with deletions detected by FISH are defined as the fraction of all investigated nuclei that have fewer locus-specific probe signals than the centromeric signals and that have at least one centromeric signal and one locus-specific probe.

A schema of deletion, amplification in interphase nuclei by dual color FISH using a combination of centromere- and locus-specific probes is shown in Figure 6, 7.

Fig. 3. FISH using a whole painting probe specific for chromosome 2. Two interphase nuclei and one metaphase spread are seen. The probe identified two copies in each nucleus. Lymphocytes. X 100

Fig. 4. FISH using a combination of abl (green) and bcl (red). Philadelphia chromosome are yellow, showing a translocation (bcr/abl) in a leukemic cell. X 1,000 Country from Ms. Sakai, Nippon Gene Research Center

Troubleshooting

- Weak or no signal
  - Specimen slides are not aged: Use one of the following: 1) Age slides for 30 mins in 2XSSC at 37°C before the denaturation procedure. Dehydrate the slides through a series of 1 minute ethanol rinses (70%, 85%, 100%), and air dry.

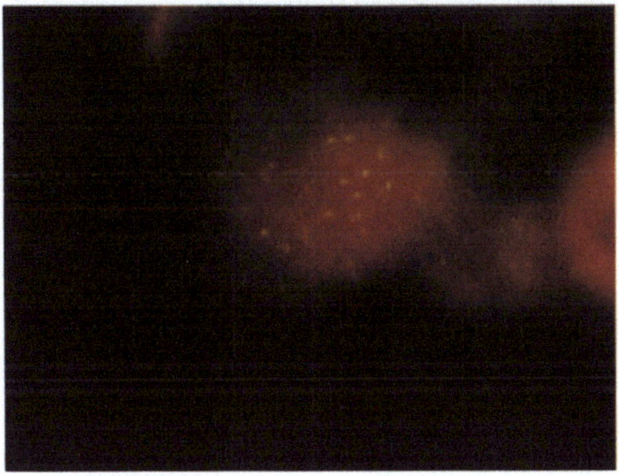

Fig. 5. FISH using an HER-2 oncogene-specific probe. A nucleus has more than 50 copies. Amplification of the oncogene may suggest the patient's poor prognosis. Ovarian cancer. Touch preparation Counterstain with propidium iodide. X 1,000

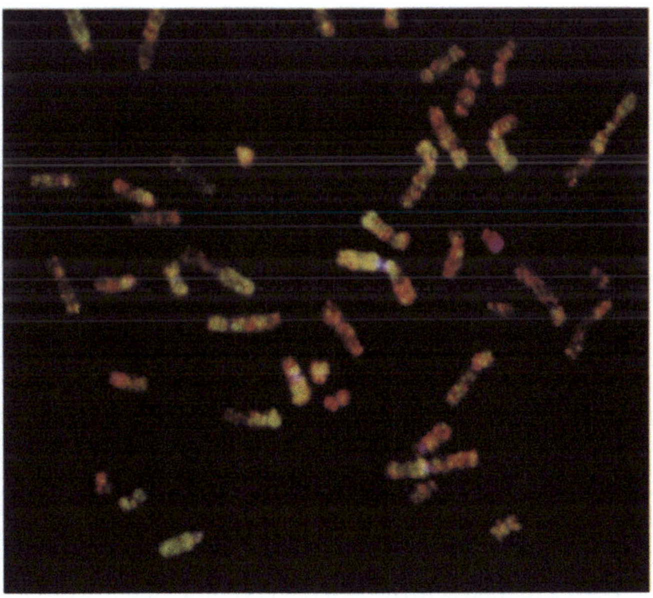

Fig. 6. Fluorescent microscopic view after CGH procedure. Cell line derived from colorectal cancer. X 1,000 Country from Dr. Uesugi, Iwate Medical University

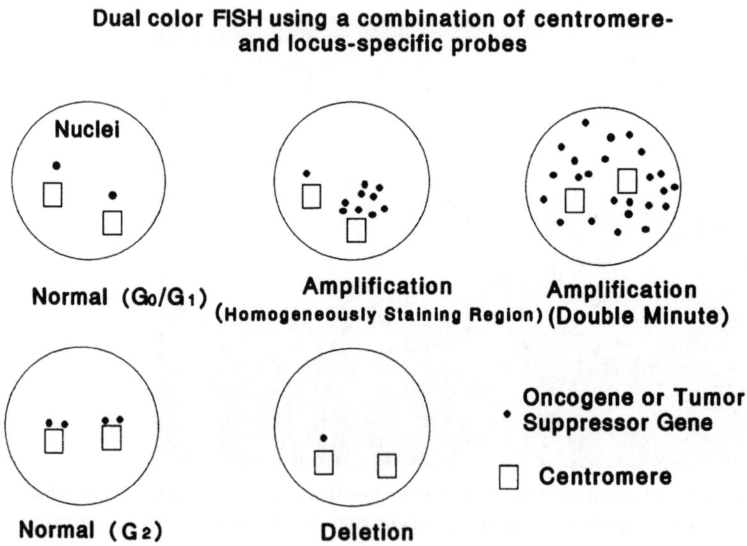

Fig. 7. Schema of deletion and amplification by dual color FISH using a combination of centromeric and locus-specific probes

2) Age slides for 24 hours at room temperature. 3) Bake slides in the air at 65°C for 5mins.

- Target DNA are not adequately denatured: Ensure temperature of denaturation solution (70% foramide/ 2XSSC) in the Coplin jar is 70°C prior to immersing the slide. Increase the denaturation time to 10 mins.
- Slides were not adequately processed for FISH: 1) Fix cell specimen in methanol-acetic acid solution (3:1). 2) Digest the slides for 7.5 mins in proteinase K at concentration 0.6 to 6 μg/ml 20 mM Tris / 2mM $CaCl_2$, pH 7.5 at 37°C, or in pepsin solution at concentration 0.01 to 0.5%/0.1N HCl. Wash in 2XSSC several times, dehydrate the slides through a series of 1 minute ethanol rinses (70%, 85%, 100%), air dry. Store slides at -20°C for up to 6 months.
- Probe was not adequately denatured: Ensure temperature of the water bath used to denature the hybridization mix is 74°C. Vortex to mix, centrifuge a few secs, and denature the hybridization mix for 5 mins.
- Air bubbles were trapped under the cover slip during hybridization: Apply cover slip by first touching the surface of

the hybridization mix. Place the slide with the cover slip down on a blotter and gently press out visible bubbles.
- Hybridization conditions are inappropriate: Ensure the temperature of the incubator is 37°C. Increase the hybridization time to 2 days. Seal the cover slip well with rubber cement.
- Wash conditions are not correct: Ensure the temperature and pH of washing solutions.
- If using hapten-modified probes, fluorescent staining procedures were not correct: Ensure the slides were adequately incubated with antibodies.
- Wrong counterstain used. Counterstain is too bright: Remove cover slip. Immerse slides for 5 mins in 2XSSC at room temperature. Dehydrate the slides through a series of 1 minute ethanol rinses (70%, 85%, 100%). Air dry and reapply counterstain.
- Viewed hybridization using inappropriate filter sets: Multi-bandpass filter sets provide less brightness than the single bandpass filter sets, probe signals may appear fainter when viewed through the multi-bandpass filter sets.

- Low signal specificity
  - Too much probe in the hybridization mix: Dilute probe 10-100 times.
  - Inappropriate hybridization conditions: Ensure temperature of incubator is 37°C.
  - Wash temperature is low: Ensure temperature of water bath is 46°C.
  - Wash solution stringency is low: Ensure the wash solutions were made correctly.

- High slide background
  - Slides were inadequately washed following hybridization: Ensure temperature and pH of wash solutions are correct.
  - If using hapten-modified probes, fluorescent staining procedures were not correct: Ensure the concentration of antibodies, often high concentration.
  - Wash solutions used too long or stored improperly: Ensure the formamide washing solution in the Coplin jars are stored at 4°C. Discard after 1 week. Discard all other wash solutions after 1 day. Ensure the washing solutions are pH 7-8.

- Distorted cell morphology
  - Specimens were not dried during preparation: Increase drying time to at least overnight at room temperature, and then age slides at least 24 hours at room temperature.
  - Specimen slides not thoroughly dry prior to immersion in denaturation solution: Dehydrate slides through a series of 1 minute ethanol washes (75%, 80%, 100%), and air dry.
  - Specimen over-digested: If using proteolytic treatment with proteinase K or pepsin, use the enzymes at lower concentration. Ensure temperature of denaturation solution is 70-74°C, and decrease the temperature to 70°C.

Subprotocol 2
Protocol for Rapid Procedure

Procedure

Note: Heat or enzymatic treatment may induce loss of adherence of cells and tissue sections to glass slides. This may be overcome by using aminoalkylsilane-coated glass slides.

Note: Prior to use, thaw reagents at ambient temperature, then centrifuge each tube for 2-3 secs using a standard bench-top microcentrifuge.

Prepare three Coplin jars: pour 50ml 100% ethanol into one jar: 50ml 85% ethanol into another, and 50ml 70% ethanol into the last.

Bring Coplin jars containing the denaturation solution and 0.4X SSC to ambient temperature. Place jars in a 74°C water bath approximately 30 mins prior to use to attain the solution temperature. Measure the temperatures of the solution inside the Coplin jar.

Hybridization mix preparation

1. Add the following to a microcentrifuge tube at ambient temperature:
   - 7µl of master mixture
   - 1µl carrier DNA (500µg/ml)

- 2μl probe (fluorochrome-directly-labeled :diluted if necessary)
- total 10μl, for each target area (18mmX18mm)

2. Centrifuge tube for 1-3 secs.
3. Vortex and re-centrifuge.
4. Place tube in a 74°C water bath for 5 mins.
5. Remove tube from water bath.
6. Place tube on a slide warmer set to 45-50°C.

Slide preparation

7. Mark hybridization areas with a diamond tipped scribe on the specimen slides.
8. Immerse the slides in the 74°C denaturation solution (70% formamide/2XSSC) for 5 mins.
9. Dehydrate slides 1 minute in 70% ethanol, 1 minute in 85% ethanol, and 1 minute in 100% ethanol.
10. Remove the slides from the 100% ethanol.
11. Dry slides by touching the bottom edge of the slides to a blotter and wiping the underside of the slides dry with a paper towel.
12. Place slides on a 45-50°C slide warmer to evaporate remaining ethanol for up to 2 mins.

Hybridization

13. Apply 10 1μl of hybridization mix to slide. Apply coverslip on placing hybridization mix.
    If hybridization time will be greater than 2 hours, seal coverslip with rubber cement.
14. Place slides in a prewarmed non-humidified box and place box in a 42°C incubator for 30 mins -60 mins.

Washing the slide

15. Remove the coverslip and immediately immerse the slide in jar of 0.4X SSC wash solution at 74°C. Agitate slide for 1-3 secs. Remove slide after 2 mins. If many slides were hybridized, do not remove the coverslips from several slides before placing any of the slides in the wash solution. Begin timing the incubation when the last slide has been added to the wash solution.

16. Wash slides in 2XSSC/0.1% NP-40 at room temperature for 1 minute.
17. Allow slide to air dry in darkness.
18. Mount in an anti-fade solution containing DAPI or propidium iodide (PI). View using a fluorescent microscope.

Subprotocol 3
Protocol for Codenaturation

Procedure

Codenaturation is a process that simplifies FISH by combining of hybridization mix and specimen into a single step. Codenaturations are performed by placing the specimen slides with probes and coverslips applied on the surface of a hot plate that is at the denaturation temperature. After 5 mins, the slides are removed and are incubated at the hybridization temperature. Conditions specify a broad range of temperatures and time, reflecting the need to optimize conditions for specific applications and specimen types.

Note: Heat or enzymatic treatment may induce loss of adherence of cells and tissue sections to glass slides. This may be overcome by using aminoalkylsilane-coated glass slides.

Preparation of a slide for co-denaturation

1. Add the following to a microcentrifuge tube at ambient temperature:
   - 7μl of master mix
   - 1μl carrier DNA (500μg/ml)
   - 2μl probe (fluorochrome-directly-labeled or hapten-labeled: diluted if necessary) total 10 μl, for each target area (18mmX18mm)
2. Centrifuge tube for 1-3 secs.
3. Vortex and re-centrifuge.
4. Mark hybridization areas with a diamond tipped pen on the specimen slide.

5. Apply 10µl ofhybridization mix to a slide and immediately apply 18 mm square coverslip.
6. Seal coverslip with rubber cement if hybridization time will be greater than 2 hours.
7. Denature the probe and specimen on the slide at 70-85°C for 5 min.
8. Hybridize at 37°C for 30 mins to 2 days.
9. Wash, fluorescent stain and counterstain: Follow the subprotocols. View using a fluorescent microscope.

Troubleshooting

The chromosome morphology observed in hybridization where codenaturation is used may differ from a specimen that is denatured and dehydrated before the probe is applied. Often, this has little effect on the analysis on interphase cells. Additionally, probe targets that are over-denatured may exhibit speckling, in which the hybridized probe appears as brightly stained spots with smaller spots radiating from the central spots. Speckling is more likely to occur when using higher intensity probes. Some speckling should not interfere with chromosome enumeration.

- Cross-hybridization
  - Repeat the hybridization on a new specimen using one of the following: 1) Increase the temperature of 0.4XSSC/0.3% NP-40 by 2°C. 2) Decrease the denaturation temperature by 2°C.
  - Probe appears dim: Repeat the hybridization on a new specimen using one of the following. 1. Increase the hybridization time. 2. Increase the denaturation temperature until the morphology becomes unacceptable. 3. Wash the slide using 0.4XSSC/0.3% NP-40 at 70-73°C.
- Diffuse signal
  - Repeat the hybridization on a new specimen using one of the following. 1.Decrease the denaturation temperature by 2°C. 2.Decrease the denaturation time. 3.Wash the slide using 0.4XSSC/0.3% NP-40 at 73-76°C.

- Poor metaphase morphology
  - Repeat the hybridization on a new specimen using one of the following.1. Decrease the hybridization temperature by 2°C. 2. Decrease the denaturation temperature. 3. Fix the slides in 4% paraformaldehyde/PB for 10 mins, wash the slides in distilled water several times, and then air-dry.

Subprotocol 4
Comparative Genomic Hybridization (CGH)

Outline

Principle of CGH (see also Figure 8):

1. Preparation of normal metaphase spreads that serve as targets for the hybridization.
2. Isolation of high-molecular-weight genomic DNA from tumor specimens.
3. Labeling of the tumor and normal DNA with different fluorochromes.
4. In situ hybridization of the labeled DNAs to a normal metaphase, washing the unbound DNA fragments.
5. Fluorescence microscopy to visualize color ratio differences along metaphase chromosomes.
6. Acquisition and display of multi-color digital images.
7. Quantitation of copy number difference by generating green to red fluorescence intensity ratio profiles for all chromosomes in a metaphase spread.
8. Combining the profiles from several metaphases to improve signal to noise ratio.
9. Interpretation of ratio changes and verification of results by control experiments.

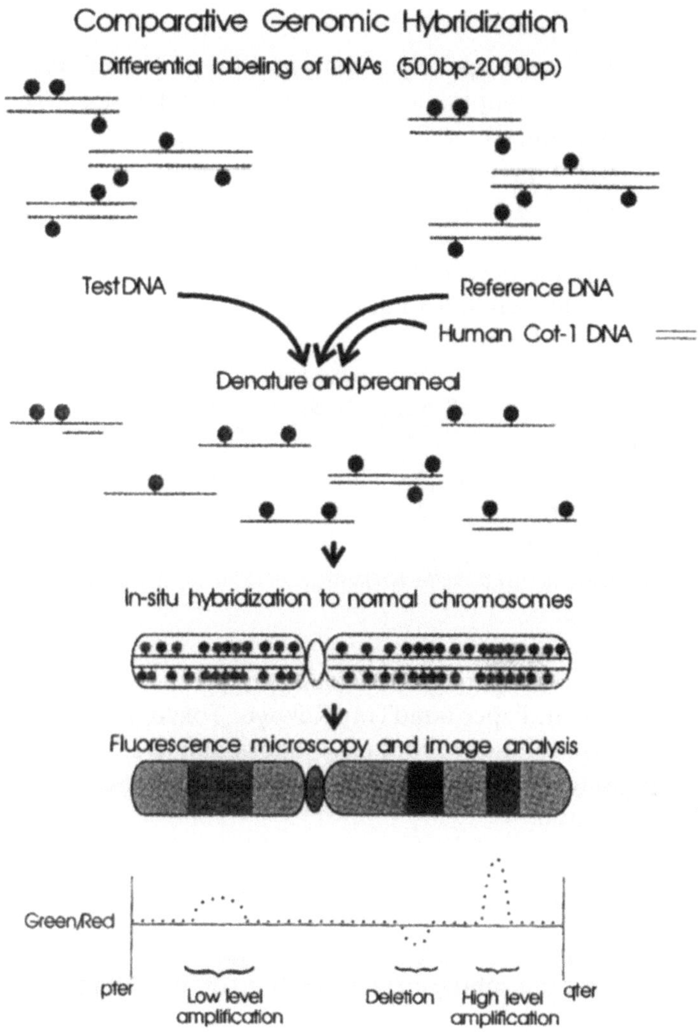

Fig. 8. Principle of comparative genomic hybridization (Houldsworth et al. 1994)

Materials

- Fluorescent microscope (e.g., Olympus BX type, Nikon) with suitable single-bandpass or multi-bandpass filter sets (for FITC, Rhodamin, Texas-Red, Spectrum-Orange, Spectrum-Green, DAPI, PI, etc.).
- Water Baths: 15°C ~ 74°C
- Vortex
- A bench-top microcentrifuge
- Microcentrifuge tubes: 0.5 ml, 1 ml
- Diamond tipped pen
- Normal male or female metaphase CGH target slides
- Glass cover slips; 18X18 mm, 22X24~60 mm
- Coplin jars
- Micropipettes (variable type:0.5~10µl, ~200µl, 1ml)
- Moist chamber
- Incubator at 37°C
- Hybridization buffer: 50% formamide/10% dextran sulfate/2XSSC.
- Purified water
- Formamide: Ultra pure grade (Kodak etc.)
- Rubber cement: Paper bondTM (Kokuyo, Tokyo, Japan)
- Denaturation solution: 50 ml of 70% formamide/2X SSC
- Washing solution: 150 ml of 50% fromamide/2X SSC
- 20X SSC solution
- PN buffer
- Antifade solution, see Recipes in Subprotocol 1(FISH), Materials; also commercially available
- DAPI or PI, 100 - 1000 ng in antifade solution
- Ethanol dehydrate solution: concentrations of 70, 85 and 100%.
- 3M Sodium acetate
- tumor DNA
- 10x A4 mixture: 10 x A4 is 0.2 mM dATP, dCTP, dGTP in 500mM Tris-HCl (pH 7.8), 50 mM $MgCl_2$, 100mM β-mercaptoethanol, 100µg/ml BSA (nuclease free).
- FITC-12-dUTP and Tex-Red-5-dUTP (DuPont, Boston, MA)
- enzyme mixture containing SNA polymerase I (~2 U) and DNAse I (~200 pg) (Gibco BRL, Gaithersburg, MO)
- DNA polymerase I (Promega, Madison, WI)
- proteinase K solution (0.1µg/ml in 20 mM Tris-HCl/2mM $CaCl_2$, pH 7.5)

- Cot-l DNA (10μg, from Gibco-BRL or Vysis)
- Normal male or female human genomic DNA

Procedure

1. Nick translation of DNA samples for comparative genomic hybridization. In a microcentrifuge tube, mix:
   - 1μg tumor DNA in 38μl double distilled water
   - 5μl 10 x A4 mixture
   - 1μl (~1nmol) FITC-12-dUTP (DuPont, Boston, MA)
   - 5μl enzyme mixture containing SNA polymerase I (~2 U) and DNAse I (~200 pg) (Gibco BRL, Gaithersburg, MO)
   - 1μl (~10 U) DNA polymerase I (Promega, Madison, WI)- For labeling of normal references DNA, FITC is substituted with Tex-Red-5-dUTP (DuPont).
2. Incubate 45-60 mins at 15°C in a water bath.
3. Stop reaction by a 10 mins incubation at 70°C

Note: Run a double-stranded 1 % agarose gel to check the probe fragment distribution. Size range should be about 500~2,000 bp. If the probe fragments are longer, increase the incubation time or the amount of enzyme mixture and if shorter, decrease the incubation time or the amount of enzyme mixture.

Comparative genomic hybridization

4. Preparation of target metaphase slides: Mark the area of the slide containing metaphases with a diamond pen (18 mm square)
5. Denature normal lymphocyte metaphase slides for 3 mins at 74°C in a denaturation solution (70% formamide/2xSSC)
6. Dehydrate in a sequence of 70, 85, and 100% ethanol, 2mins each
7. Incubate slides in a proteinase K solution (0.1μg/ml in 20 mM Tris-HCl/2mM $CaCl_2$, pH 7.5) for 7.5 mins at room temperature
8. Dehydrate as above
9. Preparation of hybridization mix:
   Mix in a microcentrifuge tube:

- 10μl labelled tumor DNA from the nick translation reaction (200 ng)
- 10μl labelled normal DNA from the nick translation reaction (200 ng)
- 10μl Cot-l DNA (10μg)
- 3μl 3M sodium acetate
- 66μl 100% ethanol

10. Precipitate by centrifuging for 30 mins in a microcentrifuge (~14,000 rpm)
11. Decant and let dry
12. Dissolve the pellet in 10μl hybridization buffer (50% formamide, 10% dextran sulfate, 2xSSC, pH 7)
13. Denature 5 mins at 70°C immediately before use.

Hybridization and washing

14. Pipette 10μl of hybridization mix on slide, apply an 18x18 mm coverslip and seal with rubber cement
15. Hybridize for two days at 37°C in a moist chamber
16. Remove coverslip and wash slides three times in a washing solution (50% formamide/2xSSC, pH 7) at 45°C for 10 mins each
17. Wash twice in 2xSSC at 45°C and once at room temperature, for 10 mins each
18. Wash once in PN buffer and once in distilled water (both 10 mins at room temperature)
19. Air-dry slides in darkness
20. Mount in an anti-fade solution containing 0.1-0.2μM DAPI
21. View the hybridization using triple-bandpass filter for DAPI, FITC, and Texas-Red.

Results

Gene amplifications 10 to 20 fold and changes including a large region may be visually detected (Fig 6). However, small deletions and other lower level changes can not be reliably determined without digital image analysis. A ratio profile of the fluorescent

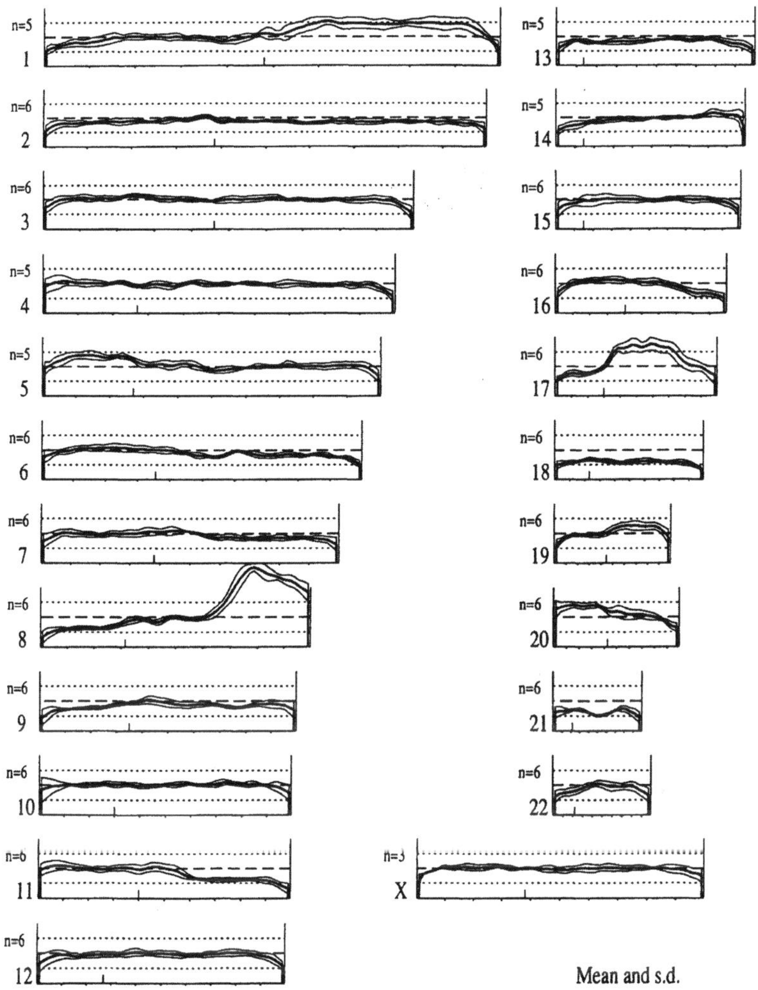

Fig. 9. Quantitative digital images of fluorescence intensity ratio (Kallioniemi et al. 1994)

intensities is measured along each chromosome; the statistical analysis of the ratios identifies regions of change in copy number (Fig. 9). However, it does not reveal translocations, aberrations, point mutations, and other aberrations that do not change copy numbers. For technical reasons, ploidy changes can not be detected, and pericentromeric, heterochromatic, and telomeric regions are ignored.

Troubleshooting

- Poor hybridization: Poor quality of metaphase slides is inadequate.
- Dim signal and/or Granular hybridizations: Probe is too small or big. Size range should be about 500~2,000 bp. If the probe fragments are longer, increase the incubation time or the amount of enzyme mixture and if shorter, decrease the incubation time or the amount of enzyme mixture. Probe concentration is low. Increase amount of probe.
- Chromosomes cannot be identified: Ensure temperature of denaturation solution is 74°C.

Acknowledgements

The principal of FISH described in this Manual was originally developed by Pinkel et al. at Laurence Livermore National Laboratory, California, USA and that of CGH by Kallioniemi et al. at University of California, San Francisco, California, USA. They have been modified at University of California, San Francisco and Tampera University, Tampera, Finland. The authors express special thanks to the technical staff at Vysis Inc. and Fujisawa Ltd. for their helpful advice, especially in Trouble shooting and Codenaturation sessions.

References

Du Manoir S, Kallioniemi OP, Lichter P et al (1995) Hardware and software requirements for quantative analysis of comparative genomic hybridization. Cytometry 19:4-9

Houldsworth J, Chaganti RSK (1994) Comparative genomic hybridization: An overview. Am J Pathol 145:1253-1260

Kallioniemi A, Kallioniemi OP, Sudar D, Rutovits D, Gray JW, Waldman FM, Pinkel D (1992) Comparative genomic hybridization for molecular cytogenetic analysis of solid tumors. Science 258:818-821

Kallioniemi OP, Kallioniemi A, Kurisu W, Thor A, Yu LC, Smith HS, Waldman FM, Pinkel D, Gray JW (1992) ERBB2 amplification in breast cancer analyzed by fluorescence in situ hybridization. Proc. Natl. Acad. Sci. USA. 89; 5321-5325

Kallioniemi A, Kallioniemi O-P, Sudar D, Rutovits D, Gray JW, Waldman FM, Pinkel D (1993) Comparative genomic hybridization: A rapid new method for detecting and mapping DNA amplification in tumors. Semin Cancer Biol 4:41-46

Kallioniemi OP, Kallioniemi A, Piper J, Isola J, Waldma, FM, Gray JW, Pinkel D (1994) Optimizing comparative genomic hybridization for analysis of DNA sequence copy number changes in solid tumors. Gene Chrom Cancer 10: 231-243

Matsuta M, Matsuta M, Hayashi S, Yasumi S, Sasaki K, Kuroda Y, Nishiya I (1996) Reconstructed Three-dimensional Images of Flow-sorted Nuclei Hybridized with Chromosome-specific Probes. J Histochm Cytochem 44: 1337-1343

Matsuta M, Matsuta M, Kon S, Thompson C, LeBoit PE, Weier HU, Gray JW (1994) Interphase cytogenetics of melanocytic neoplasmic numerical aberrations of chromosomes can be detected in interphase nuclei using antromeric DNA probes. J Cutan Pathol 21:1-6

Persons DL, Hartmann LC, Herath JF, et al. (1993) Interphase molecular cytogenetic analysis of epithelial ovarian carcinomas. Am J Path 142:733-741

Pinkel D, Gray JW, Trask B, van den Engh G, Fucoe J, van Dekken, H (1986a) Cytogenetic analysis by in situ hybridization with fluorocently labeled nucleic acid probes. Cold Spring Harbor Symp Quant Biol 51: 151-157

Pinkel D, Straume T, Gray JW (1986b) Cytogenetic analysis using quantitative, high sensitivity fluorescence in situ hybridization. Proc Natl Acad Sci USA 83: 2934-2938.

Piper J, Rutovits D, Sudar D, et al (1995) Computer image analysis of comparative genomic hybridization. Cytometry 9:10-26

Qian J, Bostwick DG, Takahashi S, et al.(1995) Chromosomal anomalies in prostatic intraepithelial neoplasia and carcinoma detected by fluorescence in situ hybridization. Cancer Res 55:5408-5414

Schroeck E, Veldman T, Padilla-Nash H et al. (1997) Spectral karyotyping refines cytogenetic diagnostics of constitutional chromosomal abnormalities. Human Genetics 101:255-262

VYSIONS Vol. 2: VYSIS Inc., 1997

White DM, Crolla JA, Ross FM (1995) Detection of minimal residual disease in childhood acute lymphoblastic leukemia using fluorescence in situ hybridization. Br J Haematol 91:1019-1024

Chapter 7

In Situ Hybridization for RNA: Radioactive DNA Probe

TAKASHI SAWAI AND MIWA UZUKI

Introduction

In situ hybridization (ISH) was first described in 1969 by Pardue and Gall. Tritium labeled nucleic acid was first used and detected via exposure to a nuclear track emulsion . Though ISH has been improved over the years, it still takes a long time to obtain fine and precise results especially when using tritium labeled probes as compared to hours of those non-radioacitve probes. The other potential labels such as ^{125}I and ^{32}P for ISH were recognized in the early 1970s. The limitations of radioactive isotopic detection for ISH were recognized from an early stage. Today in vitro biochemical labeling of purified nucleic acids is universally employed and in the future it seems that chemically synthesized oligonucleotides will be used.

The combination of radioisotopes and detection via a contact emulsion has been successfully used for many years. We have applied wide biochemical research such as in the field of developmental biology and pathology. Researchers need to be aware of the underlying principles of detection to select the best combination of radiolabeling and assay conditions to suit the particular experiment. Although ISH with the radioactive method is rather old, it still has the fundamental principles of understanding ISH. Our hybridization procedures are basically the same as those described by Hayashi et al. 1986, which is the modified Hafen' s method (Hafen et al. 1983).

Correspondence to Takashi Sawai, Iwate Medical University, Dept. of Pathology, 19-1 Uchimaru, Morioka-City, 020-8505, Japan (*phone* +81-19-651-5111 ext. 3510; *fax* +81-19-651-9246; *e-mail* tsawai@iwate-med.ac.jp)

Materials

Instruments and machines

Laboratory for radioisotope, photo darkroom, centrifugal separator(for Eppendorf tube), speed vacuum concentrator (Sarvant), water bath (30x30x50cm), small water bath (25cm diameter), aluminum cake pan (24x24x5cm) (G&S Metal Products Co., Inc., Cleveland, OH), Nensorb 20 column (Dupon), Eppendorf tube, column stand, expanded polystyrene box (5 liter volume), liquid scintillation counter, slide glass (highquality), distilled water, sliding microtome, artificial incubator, hot plate, sterilizer, Caplin jar, dark specimen box (for 25 slides), 22x26mm cover glass (Corning Glass Works, Corning, NY).

Chemical reagents

Specific cDNA probe (300bp~1kbp fragment, 20~50 λ), ^{3}H- labeled dTTP or ^{35}S- labeled dTTP (Dupon), 5x nick translation buffer (50mM Tris-HCl, pH 7.5, 5mM $MgCl_2$, 10mM ß-mercaptoethanol, 50 μg/ml BSA), unlabeled dNTP (0.3mM each of dATP, dCTP, dGTP in 50mM Tris-HCI, pH 7.6), 0.03 μg/ml DNase I (Boehringer Mannheim Biochemicals, Indianapolis, IN), Escherichia coli DNA polymerase I (Boehringer Mannheim Biochemicals, Indianapolis, IN), 0.2M EDTA, salmon sperm DNA (80μg/ml) (Funakoshi), pronase E (Type XXV; Calbiochem), formamide, paraffin for embedding, 3-aminopropyltriethoxysilane (Aldolich), 4% paraformaldehyde (EM grade, TAAB) / phosphate buffered saline (PBS) fixation buffer, ethanol (for HPLC) , xylene (for HPLC), rubber cement (Carter's Ink Co., Waltham, MA), siliconization solution (Sigma coat; Sigma), petroleum ether, Kodak NTB-2 nuclear track emulsion, Harris' alum hematoxylin (Harleco, Gibbstown, NJ), permount (Fisher Science), phosphate buffered saline (PBS), TNEB-DS mixture (20mM Tris-HCl, 0.3 M NaCl, 2mM EDTA, pH 7.0, 2x Denhardt's mixture, 20% dextran sulfate), 2M dithiothreitol (DTT) solution, methanol (for HPLC), yeast t-RNA (500 μg/ml), 0.6 M ammonium acetate, pH 7.0, Kodak D-19 developer, dextran sulfate (Pharmacia P-L Biochemicals, Piscataway, NJ), Denhardt's mixture, 20x Standard sodium citrate buffer (SSC).

Radioisotope labels

The requirements of ISH have limitations as to the radioisotope that can be used. Although it is said that the ideal radioisotope would allow a rapid result with excellent localization and the minimum of safety problem, it would be a pure ß-emitter

Table 1. Characteristics of radionuclides used in *in situ* hybridization

Radio-nuclide	T ½*	Type/max. energy of emission (MeV)	Specific activity range of nucleotides (TBq)**	Labelling methods	Typical specific activity of probe (dpm/μg)	Detection limit # (dpm/cm)
^{32}P	14.3 d	β/1.71	15-222 (400-6000)	Nick translation, random priming, in vitro transcription, end labelling	$5x10^8$, $5x 10^9$, $1.3x 10^9$, $5x 10^6$	50
^{33}P	28 d	β/0.25	15-100 (400-2500)	Nick translation, random priming, in vitro transcription, end labelling	$1x10^8$, $7x10^8$, $1.3x10^9$	300
^{35}S	87.4 d	β/0.167	15-55 (400-1500)	Nick translation, random priming, in vitro transcription, end labelling	$1x10^8$, $7x10^8$, $1.3x10^9$	400
^{125}I	60 d	β/0.035	40-80 (1000-2000)	Nick translation, random priming, direct iodination	$1x10^8$, $1.5x10^9$, $2x10^8$	100
^{3}H	12.35 y	β/0.018	0.1-4 (25-100)	Nick translation, random priming, in vitro transcription, end labelling	$5x10^7$, $1.5x10^8$	8000 ##

* Radionuclide half-life
** Values in parentheses are equivalents in curies
# Detection using an intensifying screen
Detection by fluorography using Amplify

with an energy around 0.04 MeV. However, no ideal isotope is present, so the design of the ISH procedure is a compromise between the properties of available radioisotopes, the thickness of the samples and emulsion and the type of assay result sought. Table 1 shows properties of the various isotopes that have been employed successfully for ISH (Brady and Finlan 1990). We have studied in situ hybridization using tritium labeled probes for a long time in the field of tissue destruction often seen in inflammation such as rheumatoid arthritis. Tritium is surely the low ß energy compared with ^{35}S, so exposure time required to obtain good results is fairly long, even over several weeks, compared with other kinds of radioisotopes. It gives, however, a superb resolution, as a particular definitive cellular resolution is sought. Hereafter, we describe ISH using a tritium labeled probe, especially in the field of matrix metalloproteinase and collagens.

Procedure

Labeling probe (Nick translation)

There are many kinds of probes such as cDNA, RNA and oligonucleotides and also many kinds of labeling methods, nick translation , end labeling and random labeling. Researchers should select the most suitable radioisotope, probe and labeling method, to obtain an excellent result for each study. We will introduce one labeling method called " nick translation " which we have used.

Nick translation

1. Put 100 μCi (in 100 μl 50% ethanol) of ^{3}H-dTTP (or CTP) into an Eppendorf tube
2. Remove ethanol by speed vacuum concentrator (Make several holes with a pin on the cover of the tube)
3. Add the remaining reagents (on ice) to the tube in the following order:

H_2O	X μl
Mix by Vortex for 2 mins	
DNA 1 μg	Y μl
5 x NT buffer	10 μl
dTTP	5 μl
Mix by Vortex briefly	
DNase I (1:750 (mg/ml))	4 μl
DNA polymerase I (15 units)	3 μl
Spin for 1 sec	
Total	50 μl

5 x NT (nick translation) buffer
- 0.25 M Tris-HCl, pH 7.9
- 25 mM $MgCl_2$
- 50 mM ß-mercaptoethanol
- BSA (Miles, Pentax) 250 μg/ml. Store at -20°C

dNTP
(cold deoxynucleotide phosphate mixture)
- 0.3 mM dATP, dGTP, dCTP (or dTTP, if hot dCTP is used)

DNase I
- Stock solution: 1mg/ml DNase I (Boehringer) in 0.15 M NaCl and 50% glycerol. Divide into small aliquots and store at -20 °C
- Working solution: (1:750 (mg/ml)) To 1.5 ml of 0.9% NaCl, add 2 µl of stock DNase solution. Mix very gently. Store on ice until use.

4. Incubate at 14 °C for 2 hrs (Box made of styrofoam keeps temperature constant for 2 hrs)
5. Stop reaction by adding 5 µl of 0.2M EDTA, pH 8.0 and mix
6. Separate DNA using Nensorb 20 cartridge according to procedures suggested by the manufacturer.

Note: Wash sample with 3 ml Tris-TEA-EDTA and with H_2O 1 ml x2

7. Eluate with 1 ml of 50% methanol
8. Collect first 20 drops (ca. 500 ml)
9. Count 2 µl by Scintillation counter (ideal specific activity of labeled probe (^3H: › 1X10^7, ^{35}S: › 10^8 cpm /µg DNA)
10. Add ssDNA (1mg/ml): 8 µl for 1µg cDNA (Final conc. 80 µg/ml hybridization medium)
11. Divide into 10 aliquots and lyophilize
12. Store at -20°C.

Preparation of paraffin blocks for in situ hybridization

1. Fix tissues (10mm x 10mm x 3mm) in 4% paraformaldehyde in PBS for 1 hr at room temperature
2. Wash in three changes of PBS for 5 mins each
3. Dehydrate in a graded series of ethanol (70, 80, 90, 95, 100, 100 %), each for 30 mins
4. Clear in xylene (30 mins x 3)
5. Embed in paraffin

Solutions required

1. Phosphate Buffered Saline (PBS)

0.2 M Phosphate buffer, pH 7.4	50 ml
NaCl	8.77 g
Dilute to 1000 ml	

2. 2 x PBS

0.2M Phosphate buffer, pH 7.4	10 ml
NaCl	1,75 g
Dilute to 1000 ml	

3. 10% Paraformaldehyde
Dissolve 8 g of paraformaldehyde in 80 ml of distilled water, heat the mixture to about 60°C, add 1 N NaOH drop-wise, with occasional agitation, until the mixture becomes clear. Cool to room temperature.

4. 4% Paraformaldehyde in PBS "Fixative"
Mix 80 ml of 10% paraformaldehyde (Solution 3) with 100 ml of 2 x PBS (Solution 2). Adjust pH to 7.4 and dilute to 200 ml. Store at room temperature.

Subbed slides for in situ hybridization

1. Incubate microscope slides in the following solution overnight at 60°C.

		Final Conc.
20x SSC	30 ml	3x SCC
50 x Denhardt's mixture	4 ml	1x Denhardt
H_2O	166 ml	
Total	200 ml	

Divide the solution into five Caplin jars: 40 ml per jar; Immerse slides, place molded glass cover; Cover with aluminum foil; Put in the oven at 60°C

2. Cool to room temperature
3. Drain the solution
4. Fix for 20 mins in a mixture of

100% Ethanol	3 parts (150 ml)
Acetic acid	1 part (50 ml)
Prepare 200 ml and divide into five jars	

5. Rinse in 95% ethanol 5 mins x 2
6. Dry in air
7. Store at 4°C

Solutions required

1. 20x SSC

		Final Conc.
NaCl	87.66 g	3 M
Na Citrate	44.12 g	0.3 M
H_2O up to	500 ml	
pH 7.0		

2. 50x Denhardt

Ficoll (Sigma, type 400, F-4375)	200 mg
PVP-40 (Sigma)	200 mg
Albumins from bovine serum (Sigma A-4378)	200 mg
H_2O up to	20 ml
Filter through 0.2 μm filter unit	
Divide into 5 aliquots of 4 ml	
Store at -20°C	

Cover glass for in situ hybridization

1. Wipe off cover glass with clean gauze
2. Put in fresh (or newly made) 100% ethanol overnight
3. Dry in air by placing them on a clean filter or paper towel
4. Dip cover slides into Sigma coat (kept at 4°C)
5. Dry in air
6. Heat treatment: This heat treatment should be performed prior to actual ISH

Procedure of in situ hybridization

Sectioning

1. Cut the paraffin block to 4 mm thickness
2. Mount on subbed slides
3. Dry slides thoroughly for 3 hrs on a slide warmer at 42°C
4. Store at 4°C until use

In situ hybridization - paraffin section

1. Deparaffinize by placing slides in a Caplin jar and heating in an oven at 60°C for 1 hr
2. Add fresh xylene to the hot Caplin jar and leave the jar under a ventilation hood for 30 mins with occasional shaking (for re moving melted paraffin)
3. Wash in fresh xylene (15 mins x2) and in 100% ethanol (5 mins x2)
4. Dry in air for 10 mins
5. Fix in 4% paraformaldehyde in PBS for 10 mins (attach the sections to slides firmly)
6. Wash in PBS (5 mins x3)
7. Dehydrate in 70% ethanol (5 mins x2) and in 95% ethanol (5 mins)
8. Dry in air

Prehybridization treatment

1. Cover the sample with predigested pronase* for 10 mins at room temperature

Note: Dissolve 10 mg pronase (Calbiochem) in 4 ml H_2O and incubate for 4 hrs at 37°C. Divide into 40 aliquots of 100 μl. Store at -20°C. Before use, mix one aliquot (0.25 mg/ 100 μl H_2O) with 900 μl of 50mM Tris-HCl, pH 7.6, containing 5mM EDTA.

2. Wash in PBS containing 2 mg/ml glycine (30 sec x2) and in PBS (30 secs x2) (Stop the enzymic reaction and neutralize excess aldehyde residue)
3. Fix in 4% paraformaldehyde in PBS for 20 mins
4. Wash in PBS containing 2 mg/ml glycine (5 mins x2)
5. Immerse in 0.1 M triethanolamine buffer (pH 8.0) for 5 mins
6. Immerse in a freshly prepared mixture* of 0.25% acetic anhydride in the triethanolamine buffer for 10 mins with occasional shaking.

Note: To prepare the mixture, acetic anhydride should be added in a few minutes because half-life of this mixture is 1 min. Mix within few secs before use

7. Wash in 2 x SSC (5 mins x2)
8. Dehydrate in 70% ethanol (5 mins x2) and in 95% ethanol (5 mins)
9. Dry in air (section side down)

For negative control, after washing three times with PBS, slides are reacted with RNase solution (2 mg/ml 300mM Tris-NaCl, pH 7.5) , wash in PBS again, then dehydrate in 70% ethanol (5 mins x2) and in 95% ethanol, 5 mins and dry in air.

Prepare the hybridization mixture

Probe	0.1 μg
ssDNA	8 μg
Formamide	50 μl
TNEB + DS *	50 μl
Yeast tRNA (55mg/ml)	1 μl

***TNEB**
- 20 mM Tris-HCl, pH 7.0 (0.2M pH 7.0, 5ml)
- 0.3 M NaCl (5M 3ml)
- 2 mM EDTA, pH 7.0 (0.2M Na_2EDTA. pH 7.0, 0.5 ml)
- 2x Denhardt (50x Denhardt 2ml, H_2O 8 ml)

20% dextran sulfate
- mix 2.5g dextran sulfate with approximately 3 ml of H_2O in a 15 ml Corning centrifuge tube
- dilute to 5ml with H_2O
- boil for 10 mins.
- store at -20°C until use

TNEB-DS
- TNEB, 600 μl
- 50% dextran sulfate, 400 μl

1. Mix TNEB-DS and formamide (1 : 1) thoroughly using Vortex
2. Heat at 80°C for 3 mins and chill in ice water
3. Spin for 1 sec and keep in ice water until use

Hybridization

1. Apply 20 μl of the mixture on a pretreated slide
2. Cover the sample with a sterile 22 x 22mm cover glass carefully so as not to make bubbles, and seal its edges with a 1:1 mixture of rubber cement and petroleum ether. Dry the cement.
3. Incubate slides by placing them on an aluminum cake pan floating on a water bath and cover with Styrofoam plate at 45°C for 18 to 20 hrs so as not to drop water on the slides.

Washing

1. Peel off rubber cement with fine forceps
2. 2x SSC: 1 hr at room temp. Remove cover glass

Note: Do not remove forcefully, the cover glass comes off in solution spontaneously

3. 2x SSC: 10 mins x3 at room temp.
4. 0.5x SSC*: 10 mins at 45°C *2 x SSC: 20 ml; H_2O: 60 ml
5. 0.1x SSC*: 10 mins x3 at 45°C *2 x SSC: 12.5 ml; H_2O up to 250 ml

Autoradiography

1. Prepare NTB-2 diluted 1:1 with 0.6M ammonium acetate (4.62g/100 ml, pH 7.4, Prewarm at 45°C)
2. Dip the slides in the emulsion
3. Wipe off any excessive solution on the back of slide
4. Place the slides on a cold stainless steel plate laid horizontally on crushed ice for 3 mins
5. Dry at room temperature for 1 hr
6. Place the slides in the dark box with silica
7. Wrap the box with aluminum foil
8. Place box in the refrigerator at 4°C

Development

1. Develop the slides in Kodak D-19 for 3 mins at 17°C with intermittent agitation (Exposure time, ^{3}H: 12-14 days, ^{35}S: 5-7 days)
2. Rinse for 30 secs in deionized water
3. Put in Kodak fixative for 2 mins at 18°C
4. Wash the slides (2 mins x5)
5. Dry in air
6. Put slides on the plate over crushed ice
7. Stain with the Harris' alum hematoxylin solution (diluted 1:1) for 1 min.
8. Wash in the cold water (1 min x3) (If the water is not cold, silver grains come off.)
9. Wash in PBS for 3 mins
10. Wash in cold water (2 mins x3)
11. Dry in air
12. Dehydrate in 100% ethanol
13. Clear in xylene (5 mins x2)
14. Mount with Permout (Inappropriate mount quickly loses silver grains)

Results

- mRNA associated with production of specific amino acid could be detected by expression of silver grains. Control studies are necessary for confirming the positive results using well-known probes, RNase treatment and other negative tissues or cells.
- Immunohistochemical (IHC) study should be performed, if possible, and compared with the result of in situ hybridization to obtain more information about the positive cells.
- Northern blot is also an advantage for obtaining information as to whether or not the tissues used contain enough mRNA.
- IHC and hematoxylin-eosin stain reveals information about the cells and tissues relevant to cellular dynamics, if effectively used with ISH
- RI-ISH, whether or not mRNA is expressed, shows activity and dynamics of the cells and tissues in morphogenesis

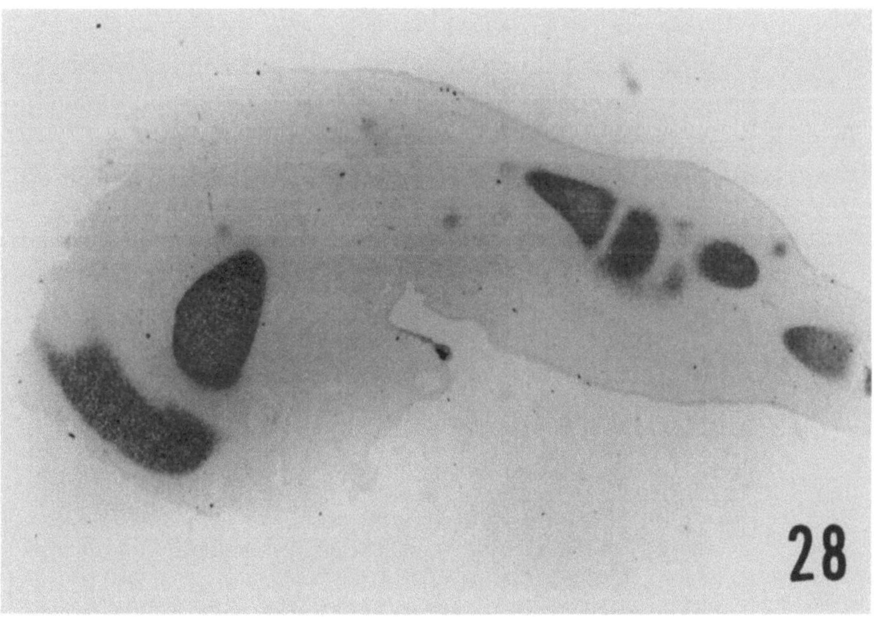

Fig. 1. mRNA of Type II collagen in the limb of chick embryo at stage 28.

and biochemistry in the field of pathology and developmental biology.

Applications

- Developmental biology
  Fig. 1 shows the mRNA of type II collagen in the limb of a chick embryo at stage 28 demonstrated by ^{3}H labeled probe in situ hybridization. Observation of the development process clearly shows that the bones grow from both peripheral regions formed in the limbs.
- Inflammation
  The left side of Fig. 2 is the MMP-1 mRNA expressed in the superficial layer of synovial membrane in rheumatoid arthritis, which plays an important role in deteriorating many kinds of protein components compromised of extracellular matrix such as type I , II , III collagens, fibronectin and laminin. On the right side in Fig. 2, the protein of MMP-1 is demonstrated by an immunohistochemical method, as far as possible immunohistochemical study should be done together.

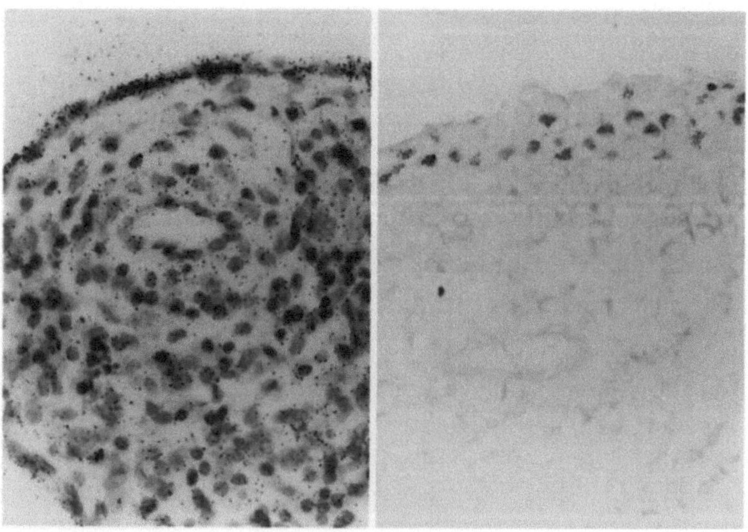

Fig. 2. mRNA (left) and enzymic protein (right) of collagenase (MMP-1) in rheumatoid synovia.

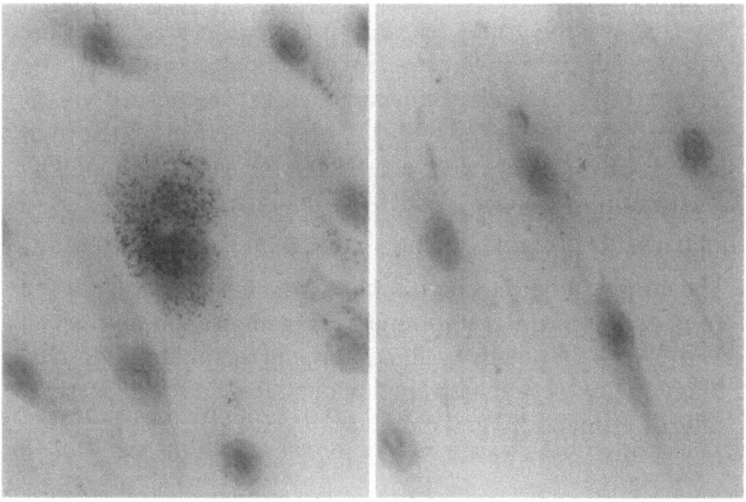

Fig. 3. mRNA of matrix metalloproteinase-1 (MMP-1), stimulated by IL-1 (left) and by γ-IFN (right).

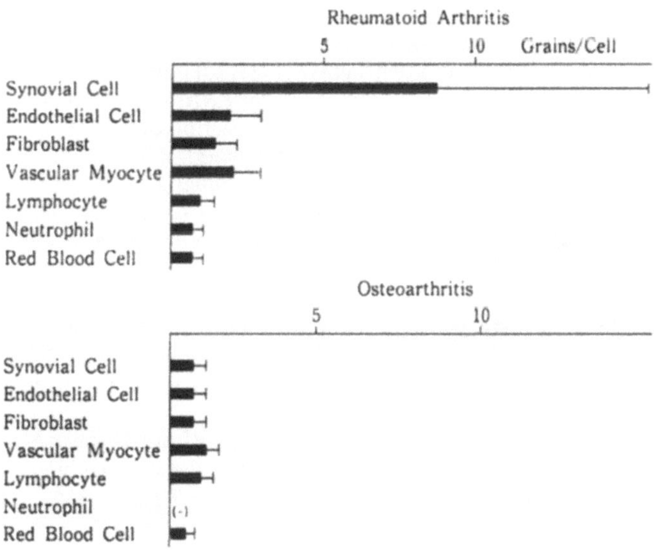

Fig. 4. Numbers of silver grains of MMP-3 among cells comprising rheumatoid and osteoarthritic synovia.

- Culture cells
  Fig. 3 shows the grains relevant to MMP-1 expressed on the culture cells of synoviocytes from rheumatoid arthritis. The left side of Fig. 3 shows the synoviocyte expressing mRNA stimulated by Interleukin-1(IL-1) , while as shown on the right side, no silver grains were detected in the cells stimulated by Interferon-γ (IFN-γ).

- Quantity of mRNA in ISH
  The number of grains is countable, which is available for the comparison of mRNA volumes. Fig. 4 shows the number of MMP-3 mRNA positive cells comprising rheumatoid synovia. Many grains were massively demonstrated in the superficial cells of synovial tissues.

References

Brady MAW and Finlan M (1990) Radioactive labels: autoradiography and choice of emulsions for in situ hybridization. In In situ hybridization - principles and practice- (eds. JM Polak and JO' D McGEE), Oxford University Press Oxford - New York - Tokyo

Hafen E, Levine M, Garber LG and Gehring WJ (1983) An improved in stiu hybridization method for the detection of cellular RNAs in Drosophilia tissue sections and its application for localizing transcripts of the homeotic Antennapedia gene complex. The EMBO J 2:617-23

Hayashi M, Ninomiya Y, Parson J et al (1986) Differential localization of mRNAs of collagen types I and II in chick fibroblasts, chondrocytes, and corneal cells by in situ hybridization using cDNA probes. J Cell Biol 102: 2302-9

Pardue ML and Gall JG (1969) Molecular hybridization of radioactive DNA to the DNA of cytological preparations. Proc. Natl Acad Sci 70: 1860-4

Sawai T and Uzuki M (1996) In situ hybridization by radioactive cDNA probes - practice and application- In Histochemistry i n 1996. ed by K Nakane, PP129-136 , Gakusai Kikaku, Tokyo

Sawai T, Uzuki M, Harris Jr. ED , Kurkinnen M, Trelstad RL and Hayashi M (1996) In situ hybridization of stromelysin mRNA in the synovial biopsies from rheumatoid arthritis. Tohoku J. Exp. Med. 178 : 315 - 330

Chapter 8

In Situ Hybridization for RNA: Radioactive RNA Probe – A Protocol for Practice With a Kit

HIDEFUMI YOSHIOKA AND SUMIHARE NOJI

Introduction

Recently, nonradioactive probe has often been used to detect RNAs in sectioned tissues by in situ hybridization (ISH) method. Although color-based detection systems with nonradioactive probe is becoming popular in ISH methods, the method with radioactive (RI) probes is still the most sensitive way to detect RNAs in the tissues. Thus, if a given mRNA cannot be detected in tissues with color-based detection methods, although it is confirmed to be present by PCR, the final way to detect the mRNA is to use the ISH method with an RI probe. Processes of specimen preparation, pre-treatments, and hybridization are similar in either radioactive or nonradioactive methods. A major difference between them is the detection method, i.e., autoradiography is used for detection of an RI probe.

An efficient and easy way to learn the methods is to use a kit including tissue sections on glass slides for a positive control, solutions necessary for ISH, probe preparation, and detection. We developed an ISH kit for practice, which is now commercially available from Nippon Gene Co.. The kit has been used successfully for various tissues at least from human, mouse, rat, and chick. In this section, the ISH kit and how to use it is introduced only in the case of using radioactive probes and paraffin sections. Additional information for using nonradioactive probes is also available in the kit.

Hidefumi Yoshioka, *Correspondence to* Sumihare Noji, The University of Tokushima, Department of Biological Science and Technology, Faculty of Engineering, 2-1 Minami-Jyosanjima-cho, Tokushima City, 770-8506, Japan (*phone* +81-88-656-7528; *fax* +81-88-656-9074; *e-mail* noji@bio.tokushima-u.ac.jp)

Outline

A protocol of the entire procedure is illustrated in Fig. 1 (Noji 1997).

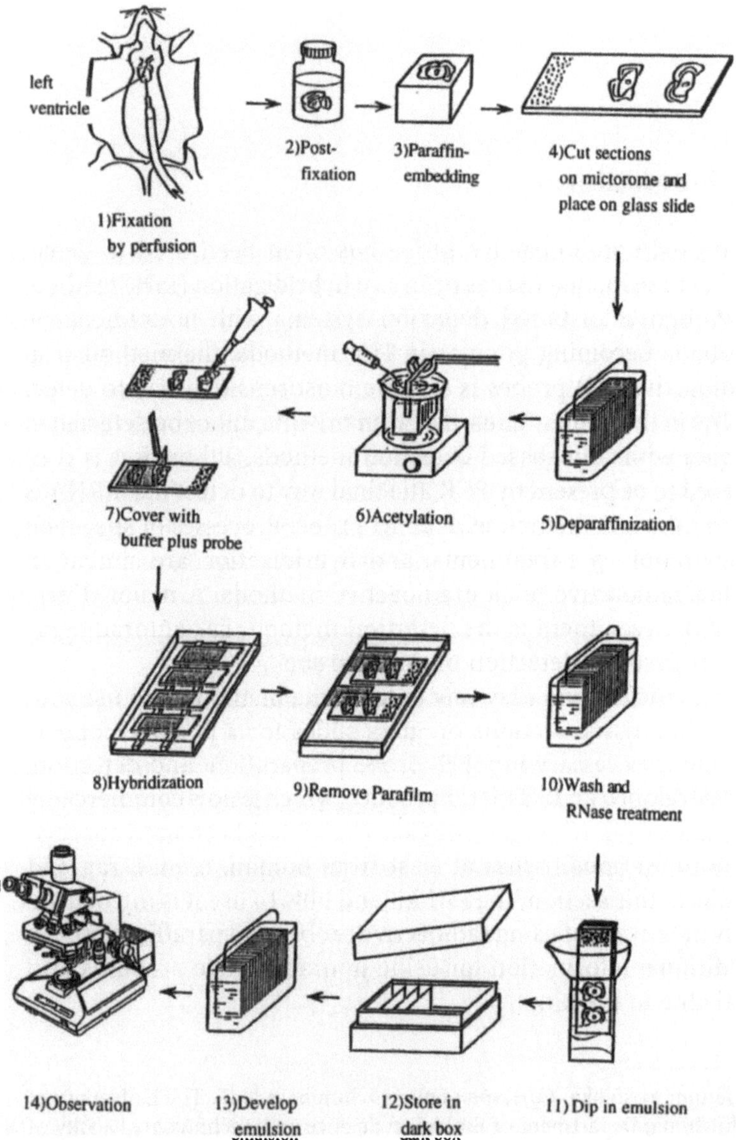

Fig. 1. Schematic illustration of procedures for in situ hybridization (Noji 1997).

Materials

Contents in the ISH kit

- Tissue sections for a positive control: 4 glass slides with sections of submandibular glands from adult male mouse (C3H).
- Solutions for preparation of RNA probe
- RNA synthesis kit
  - RNA polymerase buffer 20 µl
  - 10 mM rATP 5µl
  - 10 mM rCTP 5 µl
  - 10 mM rGTP 5 µl
  - 10 mM rUTP 5 µl
  - RNase inhibitor 5 µl
  - SP6 RNA polymerase 5 µl
  - T7 RNA polymerase 5 µl
  - T3 RNA polymerase 5 µl
  - DNase I (5 mg/ml) 5 µl
  - Sterilized water 500 µl
- Composition of RNA polymerase buffer
  - 200 mM Tris-HCl (pH 7.5)
  - 30 mM $MgCl_2$
  - 25 mM NaCl
  - 10 mM Spermidine-$(HCl)_3$
  - 50 mM DTT
  - 0.5 mg/ml BSA
- Solutions for ethanol precipitation
  Ethanchinmatc 10 µl
  - 3M Sodium acetate (pH 5.2) 100 µl
- Alkaline hydrolysis buffer 800 µl containing 40 mM $NaHCO_3$ and 60 mM Na_2CO_3 (pH 10.2).
- Linearlized DNA template for RNA probe synthesis
  - A plasmid containing a fragment of mouse nerve growth factor (NGF) cDNA
  - For antisense probe 2 µl (1 µl/ml)
  - For sense probe (as a negative control) 2 µl (1 µl/ml)
- Solutions for ISH for two sets (20 slide glasses/set)
  - ISHR1: two 600 ml PBS buffers containing 0.1 M NaCl and 10 mM sodium phosphate (pH 7.4)
  - ISHR2: 800 ml PBS/glycine buffer containing 2mg/ml glycine, 0.1 M NaCl and 10 mM sodium phosphate (pH 7.4)
  - ISHR3: 800 ml acetylation buffer containing 0.1 mM triethanolamine (pH 8.0)

- ISHR4: Two 1 ml Acetic anhydride
- ISHR5: two 1 l solutions containing 4xSSC containing 0.6 M NaCl, 0.06 M sodium citrate.
- ISHR6: Proteinase K solution (10 mg/ml) 40 μl
- ISHR7: Eight 93 μl hybridization buffers containing
  - 50% formamide, 2xSSC, 1 μg/μl tRNA, 1 μg/μl sonicated salmon DNA,
  - 1 μg/μl bovine serum albumins (BSA), and 10% dextran sulfate.
- ISHR8: 100 μl dithiothreitol (DTT) solution (1.2 M)
- ISHR9: 800 ml NTE buffer containing 0.5M NaCl, 10 mM Tris-HCl (pH 8.0), and 1 mM ethylene diaminse tetra acetate sodium salt (2NaEDTA)
- ISHR10: Two 0.4 ml solutions containing RNase A solution (10 mg/ml RNase A).
- ISHR11: Two 600 ml solutions containing 0.1xSSC, 15 mM NaCl and 15mM sodium.

Equipment and supplies to be prepared Disposable gloves, two forceps, two standard slide glass carriers (20 slots), 29 rectangular staining dishes, 500 ml beaker, a starring bar, stirrer, clamp, pipettes, 200 ml and 1 l measuring cylinder, Parafilm, incubator, rectangular culture dishes, 8 glass rods, 1 ml micropipettes, and yellow tips

Reagents Fixative (50 ml 4% paraformaldehyde):

- 10 ml of 20% paraformaldehyde (use within a month). (To prepare 20% paraformaldehyde, warm water and paraformaldehyde powder at 50-60°C and add 10N NaOH until solution becomes clear.)
- 40 ml of PB (final 0.07-0.08 M)
  - PB: solution A: 0.2 M $NaH_2PO_4 \cdot H_2O$ (27.6 g/l)
  - solution B: 0.2 M $NaH_2PO_4 \cdot 2\,H_2O$ (35.6 g/l)
  - 23 ml solution A + 77 ml solution B + 100 ml H_2O = 200 ml (0.1 M)
- PBS: 5.84 g NaCl + 1.44 g $NaH_2PO_4 \cdot 2\,H_2O$ + 0.20 g NaH_2PO_4 in 1000 ml water, 450 ml formamide, 920 ml ethanol, 200 ml xylene

Procedure

Tissue fixation and paraffin embedding

1. Write sample name, date, etc. on a small sheet of paper with pencil.
2. Prepare sample vials for fixation and put the paper in it.
3. Prepare 50 ml fixative.
4. Prepare a syringe (5 ml), a needle (27G), two forceps, and two pairs of scissors.
5. Anesthetize a mouse in a bottle containing a few ml of ethylether.
6. Wait for 2 mins and dislocate mouse neck.
7. Perfuse the mouse gently through left ventricle after cutting right atrium.
8. Remove tissues and cut the tissues in 3mm blocks.
9. Postfix the tissues overnight at 4° C in the fixative.
10. Wash with sterilized PBS three times for 15 mins to remove fixative.
11. Dehydrate by passing the tissues through a graded series of ethanol concentration for 15 mins: 50% on ice, 70%, 80%, 90%, 95%, (following processes may be performed at room temperature.),100% three times, xylene three times, xylene:paraffin 2:1 at 60° C for 1 h, paraffin for 1 h, and paraffin overnight.
12. Embed the tissues in paraffin at 62° C with the paper for identification of sample.
13. Store blocks in refrigerator.

Preparation of slide glasses

1. Put new slides in acetone containing 2% 3-aminopropyltriethoxy silane (APES) for 5 mins.
2. Wash the slides in acetone, water and ethanol for 5mins each.
3. Dry the slides at 37° C in a clean incubator.

Sectioning

1. Sterilize everything with autoclave or 70% ethanol.
2. Trim block surfaces.

3. Place in a conventional paraffin microtome.
4. Cut the paraffin block into 5 µm serial sections.
5. Transfer sections to a drop of sterilized water on an APES coated slide.
6. Dry slides on a warm plate at 42° C.
7. Store the slides at 4° C.

Preparation of RI-labeled riboprobes

1. Make water bath at 37° C for T7 and T3 or at 40° C for SP6 RNA polymerase
2. Mix these reagents in the following order at room temperature. For 5 tubes:

RNA polymerase buffer	20 µl
10 mM rATP	5 µl
10 mM rCTP	5 µl
10 mM rGTP	5 µl
10 mM rUTP	3.25 µl
[35-S]-rUTP, 14.8 -29.6 TBq (400-800 Ci /mmol)	50 µl
Sterile water	43.25 µl

3. Transfer 17 µl mixture to a 1.5 ml tube
4. Add 2 µl linearized DNA to each tube.
5. Add 1 µl RNase inhibitor
6. Add 1 µl RNA polymerase.
7. Incubate for 2 hrs at 37° C for T7 or T3 RNA polymerase or at 40 °C for SP6 RNA polymerase.
8. To check synthesis of RNA probes, remove a 1 µl aliquot and run on an agarose gel which does not need to be denatured. An RNA band should be 10-fold more intense than the plasmid band.

Incorporation ratio (optional)

a) To check incorporation of RI in probes, remove 1 µl reaction mixture to 499 µl water.

b) Spot 5 µl on two filters. One of them (Filter A) is used to measure total count, while the other filter (Filter B) is washed five times with 5 ml 10% trichloroacetic acid (TCA), followed by 5 ml of 95% ethanol.

c) Place each filter in the corresponding scintillation vial. Add 5 ml scintillator.

d) Measure radioactivity (dpm) of each vial.

e) Calculate incorporation ratio (dpm of Filter B/dpm of Filter A).

9. Add 1µl DNase I(2 U), incubate at 37 °C for 15 mins.
10. Add 1 µl 3M sodium acetate and 1 µl Ethachinmate
11. Add 2.5 vol. ethanol
12. Centrifuge 15xg for 5 mins.
13. Remove supernatant and wash with 75% ethanol
14. Dry the samples
15. Remove supernatant carefully by capillary (made by pasture pipette)

Alkaline hydrolysis of RNA probes

1. To reduce the probe size to a 150 ~ 300 bp, add 200 µl of attached alkaline hydrolysis buffer.
2. Incubate at 60° C for a period of t calculated with the following equation:
   t = (li - lf)/k li lf
   where k = 0.11 Kb/mins, li:initial, lf:final Kb.
3. Add 20 µl (1M NaOAc pH 6) and glacial acetic acid 1µl (final 0.5% v/v).
4. Add 560 µl ethanol and store -20 °C for 1 hr.
5. Spin and wash with 80% ethanol
6. Spin and remove ethanol carefully with capillary.
7. Dissolve the pellet in 25 µl of water and add 2.5 µl DTT (0.1 M) and 1 µl RNase inhibitor).
8. Store at -80° C.

Deparaffinization and pretreatment

1. Deparaffinize sections(20 slides in one carrier) by dipping sequentially for 5 mins each in: xylene three times, 100% ethanol three times, 90% ethanol, 80% ethanol, 70% ethanol, 50% ethanol, and ISHR1 for 10 mins.
2. Immerse in ISHR1 containing ISHR6 (final concentration: 1 - 10 μg/ml depending on samples)
3. Quench the proteinase activity in ISHR2 for 10 mins.
4. Wash for 3 mins in ISHR1 twice.
5. Acetylate the sections in ISHR3 by adding 1 ml of ISHR4 slowly with vigorous agitation for 15 mins.
6. Wash the slides for 10 mins in ISHR5 twice.
7. Prehybridize in 200 ml of a prehybridization solution of 50% formamide/ISHR5 containing 10 mM DTT for 30 mins at 50° C.

Hybridization

1. Prepare hybridization mixture by adding RNA probe ($5x10^4$ dpm/μl) in ISHR7.
2. Aspirate excess prehybridization mixture with a capillary tube.
3. Prepare sheets of Parafilm (2.5 cm x3.0 cm) for cover.
4. Spot 20 μl of the mixture on each slide.
5. Cover the sections with a sheet of Parafilm.
6. Place four slides on two glass rods in a rectangular plate.
7. Add prehybridization solution to humidify inside of the plate.
8. Seal the plate with vinyl tape.
9. Incubate at 50° C at least for 16 hrs.

Washing

1. Immerse the slides in 50% formamide/2xSSC containing 10 mM DTT (Use the prehybridization mixture).
2. Wait until the Parafilm floats off.
3. Transfer the slides to a slide carrier (use only for washing).

4. Immerse the holder in one of four staining dishes (prewarmed at 50° C) and shake for 20 mins. Repeat this process three times.

5. Prewarm ISHR9 at 37° C without RNase A.

6. Prewarm final washing solution of ISHR11.

7. Wash the slides in ISHR9 for RNase treatment.

8. Incubate the slides in ISHR9 after addition of ISHR10 containing RNase A at 37° C for 30 mins exactly.

9. Wash the slides with ISHR9 used in the previous step 7 for 5 mins.

10. Wash the slides in ISHR11 containing10 mM DTT for 20 mins at 50° C three times.

11. Dehydrate the samples by dipping sequentially for 3 mins each in: 70% ethanol, 90% ethanol,100% ethanol, xylene, 100% ethanol, and 100% ethanol.

12. Air dry the slides.

Autoradiography

1. Prepare silica-gel, a light-tight box, black vinyl bags, a sheet of black curtain for wrapping the box, black tape.

Note: All of the following and developing steps must be carried out in a dark room using a safe-light.

2. Prepare stock emulsion (Kodak NTB-2) by dispense into 10g aliquots and store them in plastic light-tight containers (e.g., black film container) wrapped in aluminium foil, and store at 4° C.

3. Melt stock emulsion in water bath at 42° C and gently pour 10ml prewarmed water (1:1 dilution) in the container. Mix gently by swirling to avoid making bubbles.

4. Pour emulsion in the dipping chamber immersed in the water bath at 42 °C.

5. Dip two test slides to remove the surface bubbles.

6. Immerse a pair of the slides (back to back) in emulsion at 42° C and withdraw them slowly.

7. Blot the bottom edge of the slide on a paper towel to remove excess emulsion and air dry for 20 mins in a test-tube rack.
8. Place the slide in a slide box containing silica-gel particles as desiccant.
9. Seal the box with black tape.
10. Place the box in a light-tight black plastic bag.
11. Wrap the bag in a black curtain.
12. Expose at 4° C for 14 days.

Developing

1. Transfer the slides to a slide carrier.
2. Develop the slides in Kodak D-19 developer at 20° C for 3 mins.
3. Wash in water for 1-2 mins.
4. Fix with Kodak Fixer (or Fuji fix) for 5 mins.
5. Wash the slides for 30 mins in running tap water.

Staining

1. Dip the slides in hematoxyline (Derafield) for 3 mins.
2. Wash them with running tap water for 30 mins.
3. Remove extra emulsion on the back of the slide with a razor.
4. Stain in 1% eosin for 7 mins.
5. Dehydrate the slides in a graded series of ethanol; 50%, 70%, 90%, 95%, in 100% ethanol for 15 mins three times, and then in100% Xylene for 15 mins three times
6. Place the slides on a paper towel, immediately add 150 μl Entellan New (or Permount), cover the tissues with a coverslip.
7. Dry for at least 3 hrs, placing small weights on the coverslip to force out any air bubbles.

Microscopy

1. Observe the slides with microscope equipped with both darkfield and bright field condensers. In the dark field, the silver grains become visible.

2. If successful, open a bottle of champagne within 3 days.

Photography

1. Take microscopic photograph for publication with Kodak Technical pan, or Kodak T-Max100 35 mm film.

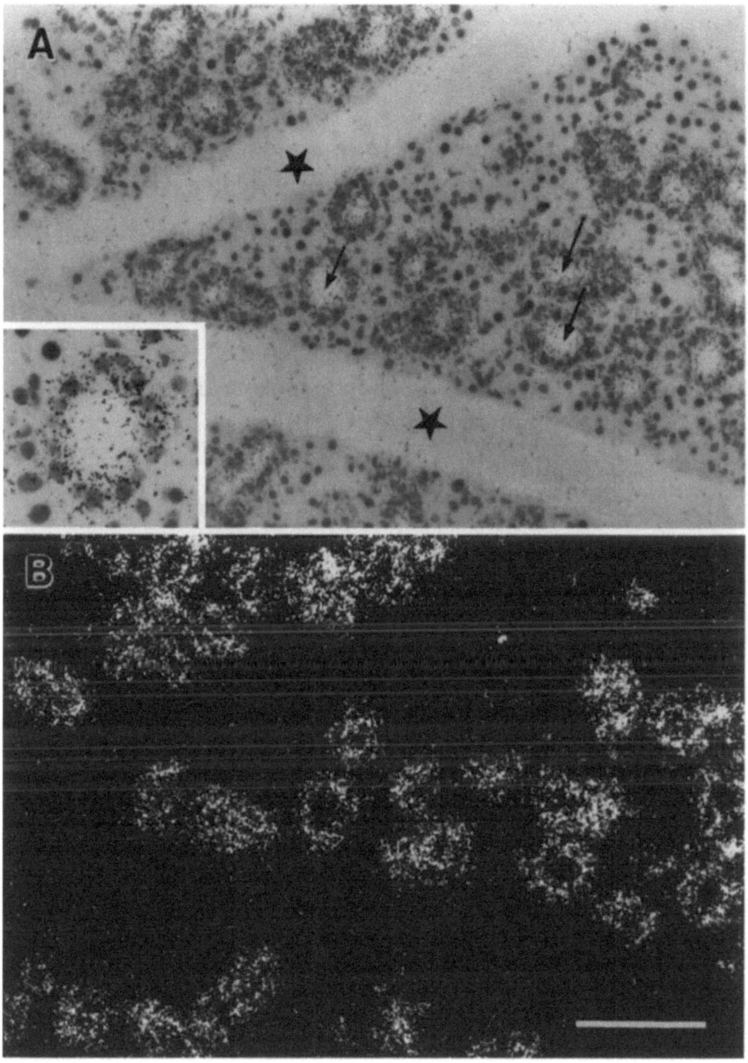

Fig. 2. Expression of EGF in mouse submandibular gland, as revealed by in situ hybridization with 35S-labeled probe. A, bright- field; B, dark-field illuminsation. Arrows indicate the granular convoluted tubules (GCT). *, interlobular connective tissues.

Results

While using this protocol, dense accumulation of signal grains indicating expression of the epidermal growth factor (EGF) gene was observed in the granular convoluted tubules of a mouse submandibular gland, as shown in Fig. 2 (Noji 1990).

Troubleshooting

- Problem 1. Sections peel off from a slide glass.
  Remedy: Check surface treatment of slide glasses.
- Problem 2. Emulsion layer peels off from a slide glass.
  Remedy: Clean the surface of the slide glasses with xylene before coating it with emulsion.
- Problem 3. RNA probe is not synthesized.
  Remedy: Check type of RNA polymerase or linearlized sites. Check purity of template DNA.
- Problem 4. High background.
  Remedy: Check probe size. Check emulsion and developing chemicals (Use fresh). Check safelights.
- Problem 5. Positive control is negative.
  Remedy: Check fixation processes. Check probes.

References

Noji S., Takahashi N., Nohno T., Koyama E., Yamaai T., Muramatsu M., and Taniguchi S (1990) Expression pattern of the homeobox gene Hox-3.5 during mouse development, as revealed by a simplified in situ hybridization method. Acta Histochem. Cytochem. 23:353-366.

Noji S. (1997) A practical use of a kit for in situ hybridization with tissue section. In: Immunostaining and In situ hybridization (Yodosha The Protocol Series). Yodosha, Tokyo, pp58-79

Angerer LM. and Angerer RC (1992) In situ hybridization to cellular RNA with radiolabelled RNA probes. In: Wilkinson DG (ed) In situ hybridization, IRL press, Oxford New York Tokyo, pp15-32

Suppliers

COMPANY: NIPPON GENE, 1-29 Tonnya-cho, Toyama City, 930-0834, Japan (*phone* +81-764-51-6548; *fax* +81-764-51-6547; *e-mail* XLD07146@niftyserve.or.jp, www.kongo.co.jp/npgene

Chapter 9

In Situ Hybridization for RNA: Nonradioactive Probe: ds cDNA Probe

HIROTAKA MATSUMI AND TETSU YANO

Introduction

As has been shown previously (Koji et al. 1988; 1990; 1996), nonradioactively labeled probes are widely used for in situ hybridization because they have several advantages compared to radioactive ones.

A double-stranded cDNA (ds cDNA) probe is the most conventional. Recently, the use of a single-stranded cDNA and a cRNA probe has been introduced. A variety of nonradioactive labels such as biotin, digoxigenin and bromodeoxyuridine are now available.

In this protocol, the thymine-thymine (T-T) dimer method is used for labeling of ds cDNA (Koji et al. 1990; Hakuno et al. 1996; Matsumi et al. 1998). In the T-T dimer method, ds cDNA, which is assumed to contain more than two adjacent thymine residues, is directly dimerized by UV-irradiation and can be used as a haptenic probe. The detection of T-T dimerized probe is carried out with immunohistochemical methods, using an anti-T-T dimer antibody.

The conditions that a probe should fulfill if used for in situ hybridization are that the labeled probes must be able to penetrate into cells and tissues and hybridize with nucleic acids, and that the hybrids must be detectable using a specific detection system. The recommended size of probe is less than 200-400 bp. Therefore, a large size of ds cDNA is used after digestion

Correspondence to Hirotaka Matsumi, University of Tokyo, Department of Obstetrics and Gynecology, Faculty of Medicine, 7-3-1, Hongo, Bunkyo-ku, Tokyo, 113, Japan (*phone* +81-3-3815-5411, Ext.3407; *fax* +81-3-3816-2017)

with a restriction endonuclease. This chapter describes the protocol for preparation of a T-T dimerized ds cDNA probe and detection of mRNA transcripts in tissue sections.

Materials

for digestion of ds cDNA probe with restriction endonucleases

Restriction endonuclease cleavage is accomplished simply by incubating the enzyme with ds cDNA in appropriate reaction conditions. The amounts of enzyme and DNA, the buffer ionic concentrations, and the temperature and duration of the reaction will vary depending upon the specific application.

- ds cDNA probe
- Restriction endonuclease
- 10x restriction endonuclease buffer
- TE buffer (pH 7.4); 10mM Tris-HCl (pH 7.4), 1mM EDTA
- 50 mM EDTA (optional)

Notes

When an appropriate enzyme cannot be found out, we use DNase I with the following buffers.

- ds cDNA probe
- DNase I (Sigma)
- 1x SSC (standard saline citrate); 0.15 M NaCl, 0.015 M sodium citrate (pH 7.0)
- 50 mM $MgCl_2$
- TE buffer (pH 7.4); 10mM Tris-HCl (pH 7.4), 1mM EDTA
- 50 mM EDTA (optional)

for T-T dimerization

- UV lamp (254 nm wave length) (Toshiba Ltd., Tokyo, Japan)
- UV radiometer (Tokyo, kougaku Kikai, Ltd., Tokyo, Japan). It is required for taking an accurate measurement of 5 $J/m^2/sec$
- Quartz dish (GL Sciences, Ltd., Tokyo, Japan)
- Forceps for handling a quartz dish

for in situ hybridization (direct method)

- Paraformaldehyde (PFA) (MERCK)
- Proteinase K (Wako); Dissolve the proteinase K at 1 mg/ml concentration in DDW. Store at - 20 °C in small aliquots.
- Mixed bed resin (Bio-Rad)
- Formamide (nacalai tesque) (for prehybridization); Formamide used for prehybridization must be of the highest grade commercially available (nacalai tesque) and be deionized with mixed bed resin just before use. (Mix 50 ml of formamide and 5 g of mixed bed resin and stir for 30 mins.)

- Formamide (Wako) (for posthybridization wash); Formamide used for posthybridization wash does not need to be deionized.
- Denhardt's solution; [0.02% BSA, 0.02% Ficoll-400 (MW=400000), 0.02% polyvinylpyrolidone (MW=360000)]
- Yeast tRNA (Sigma)
- Salmon testis DNA (Sigma); Autoclave to fragments of 100 - 300 bp and store in aliquots at - 20 °C.
- Dextran sulfate (MW=500000)
- 20x SSC (1x SSC = 0.15 M NaCl, 0.015 M sodium citrate, pH 7.0)
- Bovine serum albumin (BSA, 98-99 % pure, Sigma)
- Normal goat IgG (Sigma)
- Horseradish peroxidase (HRP)-labeled mouse anti-T-T antibody (Kyowa Medex).

Notes

- 4 % PFA/PBS; Dissolve 4 g of PFA in 80 ml of 1x PBS with heat, keeping the temperature between 55 - 60 °C. Stir the slurry until the powder is completely dissolved, and, if necessary, add a few drops of 10N NaOH to clear the solution. Adjust the volume to 100 ml and store at 4 C before use. Four % PFA/PBS is usually prepared freshly before use, but can be stored at 4 °C for up to 2 weeks.
- The constituents of hybridization mixture; 10 mM Tris/HCl buffer (pH 7.4), 0.6M NaCl, 1mM EDTA, 1 X Denhardt's solution, 250 µg/ml yeast tRNA, 125 µg/ml salmon testis DNA, 10 % dextran sulfate, 40 % deionized formamide and 0.5 - 4 µg/ml T-T dimerized ds cDNA probe. It is denatured for 10 mins before hybridization.
- The blocking solution contains 500µg/ml normal mouse IgG, 5 % BSA, 0.3 M NaCl, 100µg/ml salmon testis DNA, 100µg/ml yeast tRNA in PBS.
- The chromogen solution contains 3, 3-diaminobenzidine tetrahydrochloride (DAB), H_2O_2, nickel and cobalt; 0.5 mg/ml DAB/0.025 % $CoCl_2$ /0.02 % $NiSO_4(NH_4)_2SO_4$/ 0.01 % H_2O_2/0.1 M phosphate buffer (pH 7.5)

for in situ hybridization (indirect method)

- Paraformaldehyde (PFA) (MERCK)
- Proteinase K (Wako); Dissolve the proteinase K at 1 mg/ml concentration in DDW. Store at - 20 °C in small aliquots.
- Mixed bed resin (Bio-Rad)
- Formamide (nacalai tesque) (for prehybridization); Formamide used for prehybridization must be of the highest grade commercially available (nacalai tesque) and be deionized with mixed bed resin just before use. (Mix 50 ml of formamide and 5 g of mixed bed resin and stir for 30 mins.)
- Formamide (Wako) (for posthybridization wash); Formamide used for posthybridization wash does not need to be deionized.
- Denhardt's solution; [0.02% BSA, 0.02% Ficoll-400 (MW=400000), 0.02% polyvinylpyrolidone (MW=360000)]
- Yeast tRNA (Sigma)
- Salmon testis DNA (Sigma); Autoclave to fragments of 100 - 300 bp and store in aliquots at - 20 °C.
- Dextran sulfate (MW=500000)
- 20x SSC (1x SSC = 0.15 M NaCl, 0.015 M sodium citrate, pH 7.0)

  Bovine serum albumin (BSA, 98-99 % pure, Sigma)
- Normal goat IgG (Sigma)
- Anti-T-T mouse IgG (Kyowa Medex)
- Horseradish peroxidase (HRP)-labeled goat anti-mouse IgG (Chemicon)

Notes

- 4 % PFA/PBS; Dissolve 4 g of PFA in 80 ml of 1x PBS with heat, keeping the temperature between 55 - 60 °C. Stir the slurry until the powder is completely dissolved, and, if necessary, add a few drops of 10N NaOH to clear the solution. Adjust the volume to 100 ml and store at 4 °C before use. 4 % PFA/PBS is usually prepared freshly before use but, can be stored at 4 °C for up to 2 weeks.
- The constituents of hybridization mixture; 10 mM Tris/HCl buffer (pH 7.4), 0.6M NaCl, 1mM EDTA, 1x Denhardt's solution, 250 μg/ml yeast tRNA, 125 μg/ml salmon testis DNA, 10 % dextran sulfate, 40 % deionized formamide and 0.5 - 4 μg/ml T-T dimerized ds cDNA probe. It is denatured for 10 mins before hybridization.

- The hybridization mixture is usually prepared immediately prior to use but can be stored at - 20 °C for up to 6 months.
- The blocking solution contains 500μg/ml normal goat IgG, 5 % BSA, 0.3 M NaCl, 100μg/ml salmon testis DNA, 100μg/ml yeast tRNA in PBS.
- The blocking solution is usually prepared immediately prior to use, but can be stored at - 20 °C for up to 6 months.
- Typically, the concentration of probe is between 0.5 - 4 μg/ml. Increasing the probe concentration tends to result in higher levels of background as well as in situ hybridization signal.
- The chromogen solution contains 3, 3-diaminsobenzidine tetrahydrochloride (DAB), H_2O_2, nickel and cobalt; 0.5 mg/ml DAB/0.025 % $CoCl_2$ /0.02 % $NiSO_4(NH_4)_2SO_4$/ 0.01 % H_2O_2/0.1 M phosphate buffer (pH 7.5)

Procedure

Unless otherwise specified, all steps are performed at room temperature.

All the steps up to hybridization should be carried out in RNase free conditions.

Solutions should be autoclaved or prepared with sterile ingredients using RNase free equipment. Equipment should be autoclaved, or baked at 240 °C for 4 hrs. (Gauses should be baked at 180°C for 2 hrs.)

Digestion of ds cDNA probe with restriction endonucleases

1. Dilute the ds cDNA to a concentration of 50 μg/ml with TE buffer in a clean microcentrifuge tube.
2. Pipet the following into a clean microcentrifuge tube; ds cDNA, 10x restriction buffer and H_2O
3. Add a restriction endonuclease and incubate the reaction mixture for 10 - 30 mins at the recommended temperature (in general, 37 °C).

4. Boil the solution for 10 mins to terminate the reaction.
5. Store the digested probe at -20 °C before use.

Notes

- The restriction endonuclease is incubated in a water bath at 37 °C for 30 mins before this procedure. Because the enzyme tends to change in the activity, it is stored at -20° C in small aliquots, which is used once and discarded.
- The effect of restriction endonuclease digestion varies with the length of ds cDNA , so the time of reaction should be optimized by agarose gel electrophoresis of the digested probe. (the recommended size of probe is less than 200-400 bp).
- The volume of restriction endonuclease added should be less than 1/10 volume of the final reaction mixture, because proteins in the enzyme may interfere with the hybridization reaction.
- The digestion was also terminated by the addition of 50 mM EDTA.
- As for the control, lambda phage DNA (DNase I fragments) or pUC 18 plasmid (DNase I fragments) were used.
- To digest ds cDNA with DNase I, 10 X restriction buffer contains 1 X SSC and 50mM $MgCl_2$, and lambda phage DNA is digested for 10 mins at 37 °C.

T-T dimerization

1. Switch on UV lamps at least 15 mins before use.
2. Apply 30-60 µl of DNA solution in several drops onto the dish, and seal it with a quartz cover.
3. Irradiate it for 40 mins.

Notes

- This volume is due to the capacity of a quartz dish.
- The appropriate dose of irradiation is 12000 J/m^2 12000 J/m^2 = irradiation for 40 mins with UV lamp (5 $J/m^2/sec$)
- Immediately after use of dishes, rinse them with DDW 3 times. After rinsing, immerse them into absolute ethanol and dry up.

In situ hybridization (direct method)

1. Cut sections 5-6 μm thick and place them on silane-coated glass slides according to routine procedure.
2. Prior to use, warm at 60 °C at least 30 mins, deparaffinize through toluene and then rehydrate with serial ethanol solution and wash 3 times with PBS for 5 mins each.
3. Immerse in 0.3 % H_2O_2 in methanol for 15 mins to block endogenous peroxidase activity and wash 2 times with PBS for 5 mins each.
4. After a brief washing with DDW, immerse in 0.2 N HCl for 20 mins and wash successively with DDW.
5. Treat with 0.2 % Triton X-100 in PBS, shaking gently, for 10 mins and wash with PBS for 5 mins.
6. Treat with proteinase K (1-100 μg/ml in PBS) at 37 C for 15 mins and wash 3 times with PBS for 5 mins each.
7. Immerse in 4 % PFA/PBS for 5 mins to postfix the target nucleic acid.
8. Immerse twice in 2 mg/ml glycine/PBS for 15 mins each to remove the PFA.
9. After briefly washing with PBS and DDW, successively, immerse in 40 % deionized formamide/4 x SSC until used for hybridization (more than 30 mins).
10. Wipe each slide around the tissue and put 30-40 μl of the hybridization mixture on each slide and keep it in a moist chamber humidified with 40 % deionized formamide/4 x SSC at 37 °C for 12 hrs. Under these conditions, no covers to avoid evaporation of the medium are needed.
11. Wash 5 times with 50 % deionized formamide/0.5x SSC at 37 °C for 1hr each and twice at 45 °C for 15 mins each and twice with 2x SSC for 15 mins each.
12. After washing with PBS for 5 mins, drain each slide with a gauze, put 30-40 μl of the blocking solution on each slide and leave to stand for 1 h in a moist chamber humidified with PBS.

13. Wipe each slide around the tissue and react with 30-40 μl of the antibody (HRP-labeled mouse anti-T-T antibody) on each slide for 12 h in a moist chamber humidified with PBS.
14. Wash 4 times with 0.075 % Brij 35/PBS for 15 mins each.
15. After a brief washing with PBS, visualize the HRP sites in a chromogen solution containing DAB, H_2O_2, nickel and cobalt.
16. Dehydrate with serial up-graded ethanol solutions, cleared with xylene and mounted without any counterstaining.

Notes

- Proteinase K solution is incubated in a water bath at 37 °C for 30 mins before use. Because proteinase K tends to change in the activity, it is stored at -20 °C in small aliquots, which are used once and discarded. The effect of proteinase K digestion varies with the tissue and type of fixation, so the concentration of proteinase K should be optimized with an individual experiment.
- The slide should be wiped with gause as well as possible not to overdilute solutions. However, the slides must not be allowed to dry out.
- Salmon sperm DNA and yeast tRNA are added to inhibit DNase and RNase, respectively. Furthermore, the addition of 200U/ml heparin and/or 0.01% SDS to the hybridization mixture often decreased the nonspecific reaction of the probes.
- If necessary, in order to obtain more stringent condition for washing after hybridization, the concentration of formamide is increased to a level as high as 80 %, and the salt concentration is decreased to a level as low as that of 0.1x SSC.
- It is better to perform hybridization at less than 50 °C to preserve the morphology of tissues.
- The volume of hybridization mixture can be increased for large sections.

In situ hybridization (indirect method)

1. Cut sections to a thickness of 5-6 μm and place them on silane-coated glass slides according to the routine procedure.
2. Prior to use, warm at 60 °C for at least 30 mins, deparaffinize through toluene and then rehydrate with serial ethanol solution and wash 3 times with PBS for 5 mins each.
3. Immerse in 0.3 % H_2O_2 in methanol for 15 mins to block endogenous peroxidase activity and wash 2 times with PBS for 5 mins each.
4. After a brief washing with DDW, immerse in 0.2 N HCl for 20 mins and wash successively with DDW.
5. Treat with 0.2 % Triton X-100 in PBS, shaking gently, for 10 mins and wash with PBS for 5 mins.
6. Treat with proteinase K (1-100 μg/ml in PBS) at 37 °C for 15 mins and wash 3 times with PBS for 5 mins each.
7. Immerse in 4 % PFA/PBS for 5 mins to postfix the target nucleic acid.
8. Immerse twice in 2 mg/ml glycine/PBS for 15 mins each to remove the PFA.
9. After briefly washing with PBS and DDW, successively, immerse in 40 % deionized formamide/4x SSC until used for hybridization (more than 30 mins).
10. Wipe each slide around the tissue and put 30-40 μl of the hybridization mixture on each slide and keep it in a moist chamber humidified with 40 % deionized formamide/4x SSC at 37 °C for 12 hrs. Under these conditions, no covers to avoid evaporation of the medium are needed.
11. Wash 5 times with 50 % deionized formamide/0.5x SSC at 37 °C for 1hr each time and twice at 45 °C for 15 mins each and twice with 2x SSC for 15 mins each.
12. After washing with PBS for 5 mins, drain each slide with a gause, put 30-40 μl of the blocking solution on each slide and leave to stand for 1 h in a moist chamber humidified with PBS.

13. Wipe each slide around the tissue and react with 30-40 μl of the 1st antibody (anti-T-T mouse IgG) for 12 hrs in a moist chamber humidified with PBS.
14. Wash 4 times with 0.075 % Brij 35/PBS for 15 mins each.
15. Wipe each slide around the tissue and react with 30-40 μl of the 2nd antibody (HRP-labeled goat anti-mouse IgG) for 1 hr in a moist chamber humidified with PBS.
16. Wash 4 times with 0.075 % Brij 35/PBS for 15 mins each.
17. After a brief washing with PBS, visualize the HRP sites in a chromogen solution containing DAB, H_2O_2, nickel and cobalt.
18. Dehydrate with serial up-graded ethanol solutions, cleared with xylene and mounted without any counterstaining.

Notes

- Proteinase K solution is incubated in a water bath at 37 °C for 30 mins before use. Because proteinase K tends to change in the activity, it is stored at -20 °C in small aliquots, which are used once and discarded. The effect of proteinase K digestion varies with the tissue and type of fixation, so the concentration of proteinase K should be optimized with an individual experiment.
- Salmon sperm DNA and yeast tRNA are added to inhibit DNase and RNase, respectively. Furthermore, the addition of 200U/ml heparin and/or 0.01% SDS to the hybridization mixture often decreased the nonspecific reaction of the probes.
- If necessary, in order to obtain more stringent condition for washing after hybridization, the concentration of formamide is increased to a level as high as 80 %, the salt concentration is decreased to a level as low as that of 0.1x SSC. It is better to perform hybridization at less than 50 °C to preserve the morphology of tissues.

References

Adams JC (1981) Heavy metal intensification of DAB-based HRP reaction product. J Histochem Cytochem 29: 775

Koji T, Nakane PK (1990) Localization in situ of specific mRNA using thymine-thymine dimerized DNA probes. Sensitive and reliable non-radioactive in situ hybridization. Acta Pathol Jpn 40: 793-807

Koji T, Nakane PK (1996) Recent advances in molecular histochemical techniques: in situ hybridization and southwestern histochemistry. J Electron Microsc 45: 119-127

Koji T, Moriuchi T, Nakane PK (1988) Improved tissue preparation for in situ localization of specific mRNA using non-radioactive DNA probes: effects of protease digestion and probe size on signal detection in frozen and paraffin sections of rat pituitary glands. Acta Histochem Cytochem 21:187-200

Hakuno N, Koji T, Yano T, Kobayashi N, Tsutsumi O, Taketani Y, Nakane PK (1996) Fas/APO-1/CD95 system as a mediator of granulosa cell apoptosis in ovarian follicle atresia. Endocrinology 137: 1938-1948

Matsumi H, Yano T, Koji T, Ogura T, Tsutsumi O, Taketani Y , Esumi H (1998) Expression and Localization of Inducible Nitric Oxide Synthase in the Rat Ovary: A Possible Involvement of Nitric Oxide in the Follicular Development. Biochem Biophys Res Commun 243: 67-72

Chapter 10

In Situ Hybridization for RNA: Nonradioactive Probe: ss cDNA Probe

YOSHIO KANEMITSU AND TAKEHIKO KOJI

Introduction

The use of single-stranded cDNA (ss cDNA) is more advantageous in sensitivity than that of double-stranded cDNA (ds cDNA) or synthetic oligonucleotide as a probe for in situ hybridization (ISH) (Sugawara et al. 1990). In this chapter, we present a simple procedure for the production of nonradioactive ss cDNA probe by combining the two-step asymmetric PCR and thyminse-thyminse dimer (T-T dimer) haptenization.

Theoretically, ss cDNA probes are thought to be superior to ds cDNA probes in detection of target mRNA, because a ss cDNA probe does not contain a complementary strand which neutralizes itself. And compared with a oligonucleotide probe, a ss cDNA probe can cover such a long sequence that would otherwise require a number of oligonucleotides to cover it. A single stranded RNA probe also has such merits, but a RNA probe has a sticky nature which increases background signal, and therefore it often requires a ribonuclease treatment to reduce it. Furthermore, in practical use, the great stability of a DNA probe makes for easier handling and longer storage than that of a RNA probe.

In spite of the potential advantages of ss cDNA probes in ISH as stated above, the use of ss cDNA probes was only occasional, partly because the preparation of the probe with M13 phage vector requires great skill and much effort. However, as shown in this chapter, the application of the PCR to the production of the

Correspondence to Yoshio Kanemitsu, Kyushu University, Department of Physiology, Faculty of Medicine, Maidashi, Higashi-ku, Fukuoka, 812-8582, Japan (*phone* +92-642-6089; *fax* +92-642-6093); Takehiko Koji, Nagasaki University School of Medicine, Department of Anatomy, Sakamoto, Nagasaki, 852, Japan

probe made it much easier to prepare ss cDNA probes. We could produce a sufficient amount of ss cDNA with good reproducibility, using ordinary PCR system. If only a target mRNA sequence is known, we can synthesize the complementary ss cDNA to the arbitrary portion of the mRNA sequence by asymmetric PCR, following RT-PCR.

There are two types of asymmetric PCR; single-step method and two-step method. As shown in Results, we found that two-step asymmetric PCR was superior to single-step method. Two-step asymmetric PCR constantly produced approximately ten times more ss cDNA than ds cDNA in molarity, so the neutralizing effect of homologous strand included in the residual dsDNA on hybridization was negligible and we did not need to remove the ds cDNA.

There are a number of non-radioactive methods to label ss cDNA. Most of the methods utilize haptenic nucleic acid analogues such as digoxigenin-11-dUTP, and biotin-11-dUTP which are enzymatically incorporated during the generation of ss cDNA. However, haptenic analogues spoil the efficacy of the DNA production in some degree, or chemical labeling often has the problem of instability of reaction. Among nonradioactive ISH methods, T-T dimerization method which we originally developed is unique; T-T dimer, which is formed between adjacently located thymine molecules by UV irradiation, is used as a hapten and subsequently detected immunohistochemically by specific antibodies against T-T dimer. The haptenization is performed after PCR, so no interference to PCR occurs. T-T dimer formation is a simple physical reaction and the formed T-T dimers are chemically very stable. Thus we have practically no need to check the labeling efficacy in each case, and no purification step is required after the haptenization of DNA.

Combination of two-step asymmetric PCR and T-T dimerization enables mass production of sensitive ss cDNA probe for ISH, in a simple and reproducible way.

Outline

The entire protocol of in situ hybridization with T-T dimerized ss cDNA probe is illustrated in Fig. 1.

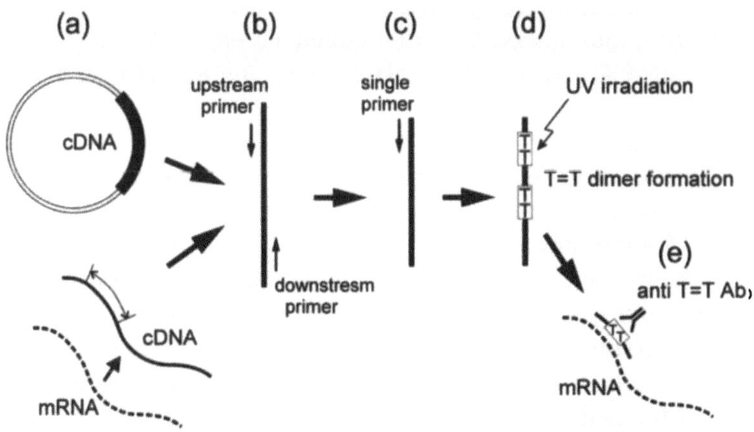

Fig. 1. An illustration of an in situ hybridization by using 'two-step' asymmetric PCR and T-T dimerization. (a) The first template for the probe production is a cloned cDNA or a reverse transcripted cDNA from mRNA. (b) The first PCR amplifies the ds cDNA using a pair of primers to produce the template for the second PCR. (c) The second asymmetric PCR generates ss cDNA using only a single primer. (d) UV irradiation forms dimers between adjacent thymines in the PCR product. (e) The haptenized cDNA is used as a hybridization probe and visualized immunohistochemically using specific antibody against T-T dimer.

Materials

- Oligonucleotide primers for PCR
- Taq DNA polymerase (e.g., Gibco BRL)
- 10x PCR buffer (Mg^{2+} free, supplied with enzyme, store at -20°C)
- 5 mM dNTP mixture (dATP, dCTP, dGTP, and TTP, 5 mM each, store at -20°C)
- Mineral oil (e.g., Sigma)
- Reverse transcriptase (e.g., M-MLV reverse transcriptase, Gibco BRL, store at -20°C)
- 5x reverse transcriptase buffer (supplied with enzyme, e.g., containing 250 mM Tris-HCl pH 8.3, 375 mM KCl, 15 mM $MgCl_2$, store at -20°C)
- 0.1M DTT
- Random hexermer (e.g., Takara, Japan, store at -20°C)
- Phenol buffered with 0.1 mM Tris-Cl/ 1 mM EDTA, pH 8.0
- 24:1 chloroform/isoamyl alcohol

- S1 endonuclease (e.g., Takara, Japan, store at -20°C)
- 10x S1 endonuclease buffer (supplied with enzyme, e.g., containing 300 mM $NaCH_3COOH$ pH 4.6, 2.8 mM NaCl, 10 mM $ZnSO_4$, store at -20°C)
- 50x TAE buffer
- Agarose for gel electrophoresis
- UV transilluminator
- Spectrophotometer
- Ultrafiltration units (cut off molecular weight 30,000 ~ 100,000, e.g., Ultrafree CL30000LTK, Millipore Co.)
- A programmable thermal controller (An conventional PCR thermalcycler is available. e.g., Model PTC-150, MJ Research)
- PCR tubes fit to the thermal controller
- Additional reagents and equipment for in situ hybridization with T-T dimer (See Chapter 11)

Procedure

Primer design

A primers pair well conditioned to ordinary symmetric PCR is probably suitable for asymmetric PCR. Requirements for the PCR primers, such as equality of melting temperatures, GC/AT balance, and careful avoidance of primer-dimer formation are also available in asymmetric PCR. One additional condition for using T-T dimerization method, is to contain the adjacent thymines to be sense probe or adjacent adenines to be antisense probe. If this condition is fulfilled, the primers which were already used to detect the mRNA, are quite suitable for usage.

Template cDNA preparation

cDNA in the vector 10^{-18} mol of purified cDNA is sufficient for 100 μl of PCR solution. It is recommended to pre-digest it by appropriate restriction enzyme to cut out the cDNA from the vector.

Reverse transcripted cDNA The first template cDNA for the probe production can be synthesized by reverse transcription from tissue mRNA. Total RNA is extracted from the tissue by ordinary guanidium thiocyanate method (Chomczynski et al. 1987) and then reverse transcripted as follows.

1. Mix the following in PCR tube:
   - Total RNA 1 μg
   - 100 μg/ml random hexermer 0.5 μl
   - DDW up to 11 μl
2. Add two drops of minseral oil and incubate 5 mins at 60°C.
3. Chill the sample tube in ice water.
4. Mix the following on ice, add to the sample under minseral oil and pipet it well:

5X reverse transcription buffer	4 μl
5 mM dNTP mix	2 μl
0.1 M DTT	2 μl
M-MLV reverse transcriptase	1 μl (200 U)

5. Set the sample tubes on the thermal controller and perform the following steps:

37°C	60 mins
95°C	5 mins
4°C	3 mins

6. Store at -20°C until usage.

Asymmetric PCR

Two-step asymmetric PCR is performed as follows, in order to produce sense or antisense ssDNAs.

First PCR

To provide ds cDNA as a template of the second PCR, the first PCR was carried out using the standard PCR conditions with a pair of primers.

1. Mix the following in PCR tube on ice:

Reverse transcription product or cDNA (10^{-18}~10^{-15} mol)	10 μl
PCR buffer (Mg^{2+} free)	10 μl
50 mM $MgCl_2$	3.4 μl* /4.0 μl
10 μM sense primer	5 μl
10 μM antisense primer	5 μl
5 mM dNTP mix	4 μl
Taq polymerase	0.5 μl (=2.5 U)
DDW	up to 100 μl

2. Set the sample tubes on the thermal controller and perform the following steps:

1 cycle:	94°C	1 mins
30 ~40 cycles:	94°C	40 secs (denaturation)
	55°C	40 secs (annealing)
	72°C	40 secs (extension)
1 cycle:	72°C	3 mins
	4°C	continuous

Annealing temperature must be changed according to the melting temperature of the primers. The optimum time parameter may differ in each case on account of the performance of the thermal controller, the volume of the solution or the sequence length to be amplified.

3. Check the size of the PCR products by agarose gel electrophoresis.

4. Remove the mineral oil. Add equal volume of buffered phenol and vortex briefly. Centrifuge 5 mins at 5000 rpm and save upper aqueous phase.

5. Add equal volume of 24:1 chloroform /isoamyl alcohol and vortex briefly. Centrifuge 5 mins at 5000 rpm and save upper aqueous phase.
6. Add 1/10 volume of 3 M sodium acetate, pH 5.5, and 2.5 volume of 100% ethanol. Mix and precipitate overnight at -20°C or 15 mins at -70°C.
7. Centrifuge 15 mins at 12000 rpm, and discard supernatant.
8. Dry pellet briefly and resuspend in 1 ml of water.
9. Put the solution in an ultrafiltration unit and centrifuge at 3000 G until 80% is filtrated. Add 800 μl of water, and centrifuge again. Repeat the centrifugation at least five times in all. Place still the last remnant at 4°C for more than 1 hr.
10. Measure the optical density of the product by spectrophotometer.

Second PCR

The second PCR amplifies only sense or antisense strand. One PCR tube (100 μl) generates 2~3 μg of ss cDNA. In order to generate sufficient amount of ss cDNA, prepare some tubes at once and react them simultaneously.

1. Mix the following in PCR tube on ice:

ds cDNA from first PCR	3 pmol (approximately 1 ng /base)
PCR buffer (Mg^{2+} free)	10 μl
50 mM $MgCl_2$	4.0 μl
10 μM primer	5 μl
5 mM dNTP mix	4 μl
Taq polymerase	0.5 μl (2.5 U)
DDW	up to 100 μl

2. Set the sample tubes on the thermal controller and perform the following steps:

1 cycle:	94°C	1 mins
40~50 cycles:	94°C	40 secs
	55°C	40 secs
	72°C	40 secs
1 cycle:	72°C	3 mins
	4°C	continuous

3. Check the PCR products by electrophoresis.

4. Correct the solutions in one tube, add same volume of buffered phenol and vortex briefly. Centrifuge 5 mins at 5000 rpm and save upper aqueous phase.

5. - **10.** is same as the first PCR.

Alternate protocol

It is possible to produce ssDNA by single step asymmetric PCR using a pair of primers, one of which is added in vast excess over the other. The concentration of one primer is from 5 to 10 nM, while that of the other is 500 nM. Other conditions of PCR are the same as the first PCR of two-step asymmetric PCR.

T-T dimerization of DNA Dilute the ss cDNA solution to 50 μg/ml and put it in a flat siliconized quartz dish with a light path of 0.5 mm. Irradiate it by a UV lamp (254 nm) at 5.0 J/m^{2}sec for 2000 secs. (See Chapter 11)

DNA fragmentation by S1 endonuclease This step may be abbreviated if the permeability of the probe is sufficient, for example, in case the probe length is shorter than 250 bases. Otherwise, digest the probe with S1 nuclease in order to reduce the size.

1. Mix the following:

50 μg/ml ss cDNA probe	35 μl
10x S1 nuclease buffer	4 μl
200 nU/l S1 nuclease	1 μl

2. Incubate 10 mins at 37°C
3. Incubate 10 mins in boiling water.
4. Confirm that the sizes of DNA fragment are about 100~300 bases by agarose gel electrophoresis.

In situ hybridization

Here I present an example of one of the possible protocols. See Chapter 11, for further explanation.

1. Place the frozen sections (4~10 um in thickness) on silane-coated glass slides.
2. Dry quickly using hair-dryer with cool air.
3. Fixed with 4% PFA in PBS for 5 mins at RT, wash in PBS 5 mins and wash in DDW 5 mins.
4. Dip in 0.2N HCl for 20 mins RT and wash in PBS 5 mins two times.
5. Dip in 1 μg/ml proteinase K for 15 mins 37°C and wash in PBS 5 mins two times.
6. Dip in 40% formamide in 4X SSC RT 30 mins.
7. Mix the following as hybridization solution (100 μl):

1 M Tris-HCl	1 μl
5 M NaCl	12 μl
200 mM EDTA	0.5 μl
100X Denhardt'	1 μl
5 mg/ml ssDNA	2 μl
10 mg/ml tRNA	2 μl
DDW	39 μl
Deionized formamide	40 μl
44 μg/ml ss cDNA probe	2.3 μl

8. Heat up the hybridization solution in boiling water for 3 mins, and chill in ice water.
9. Put the hybridization solution on the sections.

10. Incubate in moist chamber overnight at 40°C.
11. Wash in 50% formamide in 2x SSC 1hr 40°C, 5 times.
12. Wash in PBS 10 mins RT, twice.
13. Put the blocking solution (0.5 mg/ml normal mouse IgG and 5% BSA in PBS) on the section and incubate 1 hr RT in moist chamber.
14. Remove the blocking solution and put the anti-TT solution (HRP conjugated mouse anti-(T-T dimer) antibody in blocking solution) on the section. Incubate 2 hrs RT in moist chamber.
15. Wash in PBS 10 mins RT 5 times.
16. Visualize the signals in the DAB solution (DAB, nickel and cobalt in phosphate buffer).

Results

Evaluation of the PCR product

Typical results of the analysis of asymmetric PCR product are shown in Figure 2. It shows that two-step asymmetric PCR generates five times more ssDNA (second peak) than dsDNA (first peak) in weight. (It equals ten times in molarity.) Single-step asymmetric PCR produces smaller amounts of ssDNA and the ssDNA/dsDNA ratio is lower.

Gel electrophoresis of the second PCR product is shown in Figure 3. ssDNA runs relatively slower than ds DNA and its band is not sharp because of its complex second structure. The intensity of the band of ss DNA is relatively obscure when it is stained with ethidium bromide which visualizes the double stranded part. The quantity of ss DNA product increases not exponentially but linearly, and it often increases over 50 cycles.

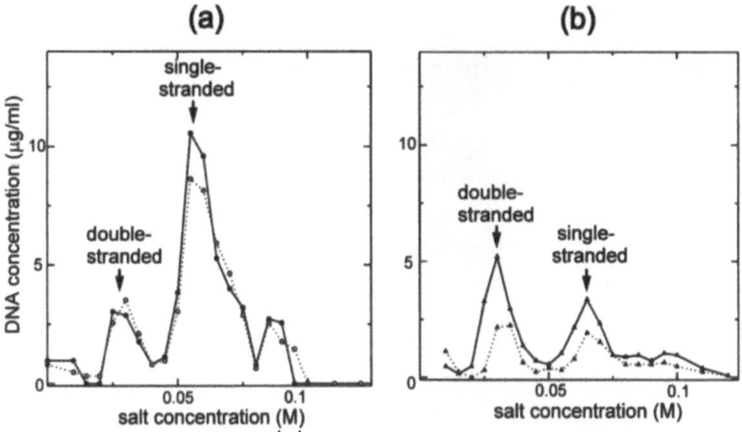

Fig. 2. Analysis of the final products of two-step (a) and single-step (b) asymmetric PCR on hydroxyapatite column. The fractions of the major peaks were evaluated by electrophoresis and filter hybridization. dsDNA (first peak) and ssDNA (second peak) were eluted separately.

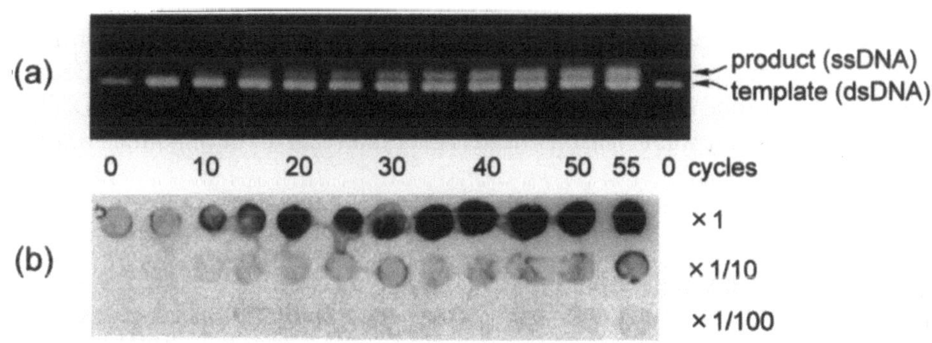

Fig. 3. Electrophoresis and filter hybridization of the second PCR product. 55 cycles of the PCR were performed and the product was sampled every 5 cycles. The samples were electrophoresed (a), or applied to the membrane and hybridized with the complementary oligonucleotide probe (b). The PCR products increased over 50 cycles.

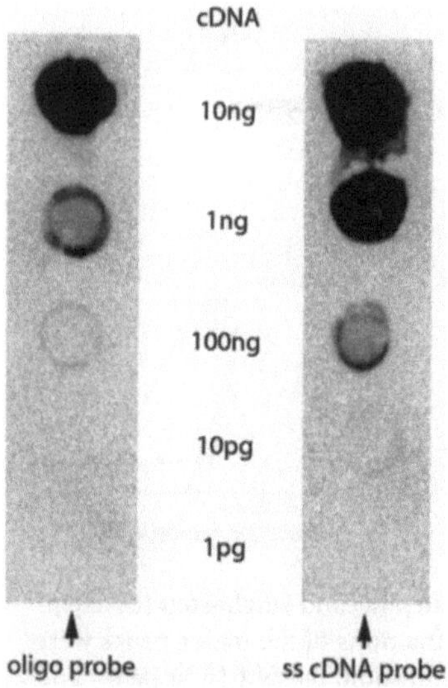

Fig. 4. Comparison of oligonucleotide probe and ss cDNA probe. Dilutions of non-labeled cDNA were applied to the membrane, and hybridized with T-T dimerized oligonucleotide probe and ss cDNA probe.

We evaluated the sensitivity of T-T dimerized ss cDNA probe and compared it with oligonucleotide probe (Figure 4). ss cDNA probe detected 10 pg of cDNA and the sensitivity of the ssDNA probe was at least ten-fold higher than that of the oligonucleotide probe.

In situ hybridization

An example of in situ hybridization using T-T dimerized ss cDNA probe is shown in Figure 5. IL-1ßmRNA in LPS stimulated lymphocytes was clearly visualized by ss cDNA probe, while oligonucleotide probe couldn't detect it clearly.

LPS(-)

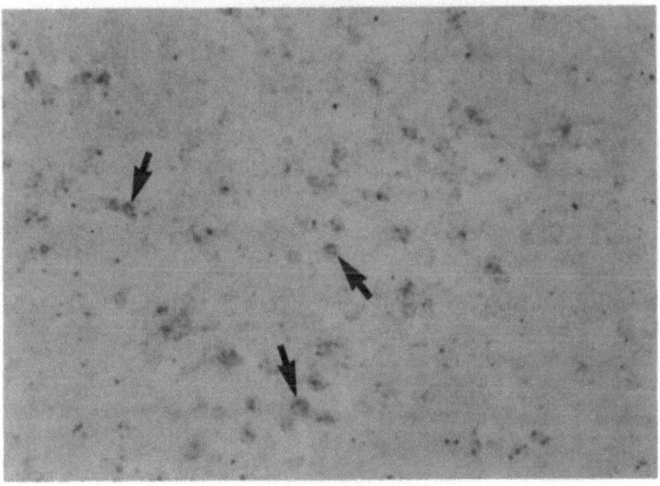

LPS(+)

Fig. 5. IL-1β mRNA localization using T-T dimerized antisense ss cDNA probe. Lymphocytes in the frozen sections of the rat spleen were hybridized with probe before (left) or after (right) LPS injection. Arrows indicate the localization in cytoplasm of positive cells.

References

Chomczynski P and Sacchi N (1987) Single-step method of RNA isolation by acid guanidium thiocyanate-phenol-chloroform extraction. Analytical Biochemistry 162: 156-159

Kanemitsu Y Koji T Hori T and Nakane PK (1994) Production of antisense DNA probe by asymmetric PCR and its application for in situ hybridization. Nucleic Acids Symposium Series 31: 57-59

Koji T and Nakane PK (1990) Localization in situ of specific mRNA using thymine-thymine dimerized DNA probes. Sensitive and reliable non-radioactive in situ hybridization. Acta Pathol Jpn 40: 793-807

Sugawara I Koji T Ueda K Pastan I Gottesman MM Nakane PK and Mori S (1990) In situ localization of the human multidrug-resistance gene mRNA using thyminse-thyminse dimerized single-stranded cDNA. Jpn J Cancer Res 81: 949-955

Chapter 11

In Situ Hybridization for RNA: Nonradioactive Probe: Oligo-DNA Probe: T-T Dimer

YOSHITAKA HISHIKAWA, MASAMOTO MURAKAMI,
MASAO KASAHARA AND NAOFUMI NAGASUE

Introduction

The method using thymine-thymine (T-T) dimerized DNA as a haptenic probe was previously reported (Koji 1988, 1989, 1990, 1996) and we have elaborated it in the following sections.

When ultraviolet rays (UV: 254nm) are exposed to the location adjacent to thymidines in the DNA strand, Pyrimidine ring of thymidine is detached and then new skeletal structure between both thymidines is constructed. This is called T-T dimer, which is very stable and has antigenic property. Using this method takes advantage of the fact that T-T dimer formation in DNA is proportional to the dose of UV irradiation (Franklin 1982) and the frequency of occurrence of two successive thymine bases in the DNA, and that no purification step for isolating T-T dimerized DNA is required. Moreover, the T-T dimer method has been revealed to have high sensitivity, good reproducibility and wide applicability.

Yoshitaka Hishikawa, Nagasaki University School of Medicine, Department of Histology and Cell Biology, Nagasaki, 852-8523, Japan (*phone* +81-95-849-7027; *fax* +81-95-849-7028); *Correspondence to* Masamoto Murakami, Fujita Health University School of Medicine, Department of Pathology, Aichi 470-1192, Toyoake-cityJapan (*phone* +81-562-93-2440; *fax* +81-562-93-2440); Masao Kasahara, Fujita Health University School of Medicine, Department of Pathology, Aichi 470-1192, Toyoake-city, Japan (*phone* +81-562-93-2440; *fax* +81-562-93-2440); Naofumi Nagasue, Shimane Medical University, Second Department of Surgery, Izumo, 693-8501, Japan (*phone* +81-853-20-2232; *fax* +81-853-20-2229)

Before starting the procedure of in situ hybridization with T-T dimer method on tissue sections, it is necessary to label the oligonucleotide with T-T dimer and then to check the labeled probe by immunohistochemical detection and the hybridizablity by dot blot hybridization on a nitrocellulose filter respectively.

Outline

- T-T dimerization of DNA
- Immunohistochemical detection of T-T dimerized DNA on a nitrocellulose filter
- Dot blot hybridization with T-T dimerized DNA probes
- In situ hybridization with T-T dimerized probes

A short protocol of the entire procedure is shown in the following:

Pretreatment Steps

- Inhibiting the activity of endogenous peroxidase
- Removing proteins such as histone and cellular matrix proteins
- Post-fixation of the specimens

Hybridization

- Base-pairing between target nucleic acids in cells or tissues and probe nucleic acids

Stringent Washing

- Reducing the level of background staining and bringing about an increase in the signal/noise ratio

Visualization System

- Direct/indirect enzyme immunohistochemistry

In detail, the following steps have to be performed:
(For specimens prepared with cross-linking fixatives such as formarin.)

Pretreatment steps

1. To inhibit the activity of endogenous peroxidase : Immerse in 0.3-3 % H_2O_2 in methanol.
2. To remove interfering proteins, especially basic proteins such as histone and RNase : Treat with 0.2 N HCl.
3. To increase the probe access to target mRNAs in tissues and to remove cellular matrix proteins : Use of proteinase K.

Hybridization

- Considering of the melting temperature (Tm) : optimal hybridization temperature = Tm - 25 °C. Use of formamide (deionized) : Deionized formamide should be used to reduce Tm. Ten percent (v/v) formamide decreases the Tm by about 6 °C (Meinkoth 1984).

Stringent washing

- This step for washing is as important as that for pretreatment steps and hybridization.

Enzyme immunohistochemistry

- Routine direct/indirect enzyme immunohistochemistry is performed.

Evaluation of results

- To confirm the specificity of target mRNA signals, various types of control experiments should be conducted concurrently.

Evaluation of results

Positive control	28S rRNA complementary probe
Negative control	Sense probe
RNase treatment	RNase-A
Competitive examinsation	Excess amount of either homologous or heterologous unlabeled oligo-DNA

Sense probe should be used as a negative control, and we use 28S rRNA probe as a positive control to evaluate the levels of hybridizable RNAs in tissue sections in every run.

Materials

for T-T dimerization

- Oligo-deoxynucleotide (Oligo-DNA)
  Oligo-DNA probe : In the case of the T-T dimer method, two or three TTA repeats are added to both the 5' and 3' ends of the original sequence and later dimerized by UV irradiation.

Design of Oligo-DNA probe

5'-TTATTA-Oligo-DNA (45 mer)-ATTATTATT-3'

    **   **   **   **   **

* = T-T dimerizable regions by UV irradiation

- Melting temperature (Tm) : $Tm = 81.5 + 16.61\log[Na^+] + 0.41(\%GC) - 675/L$ $[Na^+]$ = the concentration of sodium cation ; %GC = (guanine + cytosine)/all bases ; L = the probe length (bases)

- UV lamp (254 nm wave length) (Toshiba GL-15 ; Toshiba Ltd., Tokyo, Japan) UV radiometer (taking an accurate measurement of 5 $J/m^2/sec$) : UVR-254 (Tokyo, kougaku Kikai, Ltd., Tokyo, Japan)

Note: For the UV source, we used two UV 15 W Toshiba GL-15 (Toshiba Ltd., Tokyo, Japan) lamps placed 36.1 cm from the surface of the quartz dish. Instead of UV lamp and radiometer, uv-crossrinker (UV-stratalinker, stratagene cloning systems) is available.

- flat siliconized quartz dish : Type AB, Gaskuro Kogyo Inc. Tokyo, Japan

for immunohistochemical detection of T-T dimerized DNA on a nitrocellulose filter

- Nitrocellulose filter
- Vinyl bag for hybridization
- Gel stain Rotator
- Dry oven : Dry oven DS-44 / yamato, Japan
- Blocking solution
  - 500 µg/ml normal goat IgG : Goat IgG, purified immunoglobulin (I-5256, Sigma)
  - 5% BSA (98-99% pure) : albumins, bovine (98-99% alb) (A-7030, Sigma)
  - 100 µg/ml yeast tRNA : RNA transfer, type X from Bakers yeast (R-9001, Sigma)
  - 100 µg/ml salmon sperm DNA : salmon testes DNA for hybridization (D-7656, Sigma)
  - NaN_3
  - PBS
- 1st antibody
  - mouse anti-T-T IgG (1: 200)
  - 5% BSA (98-99% pure) : albumins,bovine (98-99% alb)
  - 100 µg/ml yeast tRNA : RNA transfer, type X from Bakers yeast
  - 100 µg/ml salmon sperm DNA : salmon testes DNA for hybridization
  - 0.05% NaN_3
  - PBS
- 2nd antibody

  HRP-goat anti-mouse IgG
  - 5% BSA (98-99% pure) : albumins, bovine (98-99% alb)
  - 100 µg/ml yeast tRNA : RNA transfer,type X from Bakers yeast
  - 100 µg/ml salmon sperm DNA : salmon testes DNA for hybridization
  - 0.05% NaN_3
  - PBS
- DAB/H_2O_2/Co/Ni solution
  - 1M PB : 1M Na_2HPO_4 and 1M NaH_2PO_4 (with DDW)
  - 1% $CoCl_2$
  - 1% $NiSO_4$-NH_4SO_4
  - 31% H_2O_2
  - DAB

for dot blot hybridization with T-T dimerized DNA probes

- Gel stain Rotator
- Dry oven : Dry oven (DS-44 / yamato, Japan)
- Constant temperature oven (DN-43H, KAYAGAKI IRIKA KOGYO,LTD., Japan)
- Water bath : constant temperature water bath (BK-43, yamato, Japan)
- Boiling bath : hot plate (type 1000, TECMO, Japan)
- Blocking solution
  - 500 μg/ml normal goat IgG : Goat IgG, purified immunoglobulin (I-5256, Sigma)
  - 5% BSA (98-99% pure) : albumins, bovine (98-99% alb) (A-7030, Sigma)
  - 100 μg/ml yeast tRNA : RNA transfer,type X from Bakers yeast (R-9001, Sigma)
  - 100 μg/ml salmon sperm DNA : salmon testes DNA for hybridization (D-7656, Sigma)
  - 0.05% NaN_3
  - PBS

for in situ hybridization with T-T dimerized probes

- the constituents of hybridization mixture

Hybridization mixture

	Volume (μl)
1 M Tris-HCl (pH 7.4)	1
5 M NaCl	12
200 mM EDTA	0.5
100 X Denhardt's solution	1
10 mg/ml yeast tRNA	2.5
10 mg/ml salmon testicular DNA	1.25
50% (w/v) dextran sulfate	20
Deionized formamide	40
50 μg/ml T-T dimerized oligo-DNA probe	4
final volume	100

If necessary, 200 U/ml heparin and 0.01% (w/v) SDS are included. Heparin is added partly to block nonspecific binding of the probes to proteins (Feavers 1989) and partly as a possible RNase inhibitor (Singh 1984). SDS decreases nonspecific staining with the sense probe.

- the constituents of blocking solution

Blocking solution
5% BSA in PBS
500 µg/ml normal mouse IgG
100 µg/ml yeast tRNA
100 µg/ml salmon testicular DNA
0.3 M NaCl

Note. Salmon sperm DNA and yeast tRNA are added to hybridization medium, blocking solution, and antibody solution as competitive inhibitors of DNase and RNase.

- HRP-labeled anti-(T-T dimer) monoclonal antibody (Kyowa Medex)

for accessories

- Triton X-100 (Sigma)
- Proteinase K (Sigma)
- PFA/PBS (PFA; MERIK)
- 2mg/ml glycine/PBS (glycine; Wako)
- deionized formamide (formamide; Nakarai)

Procedure

T-T dimerization of oligo-DNA probes

Oligo-DNA probes are required about 45-base sequence complementary to the target mRNAs. To obtain the maximum signal by in situ hybridization, it should be verified whether the probes are optimally labeled. Too little labeling results in a low signal, whereas, excessive labeling will reduce the efficiency of hybridization. Antisense and sense sequences are used for synthetic oligo-DNA probes but for haptenization of the oligo-DNAs with T-T dimers, two and three TTA repeats sequences are added at the 5'- and 3'-ends of the sequences, respectively at the time of synthesis. Recently, it has been possible to synthesize the oligo-DNAs with an automatic DNA synthesizer.

1. Find quartz dishes, two forceps and a sheet of Parafilm.
2. Switch on UV lamps at least 15-30 mins before use.
3. Check and adjust position of the stand to give a dose of 5J/ m^2/sec by using UV-radiometer.
4. Wash a quartz dish with sterile DDW before use.
5. Apply 60 μl of DNA solution (50 μg/ml)(depending upon the capacity of a quartz dish) in several drops onto the dish and seal it with a quartz cover.

Note: Be careful not to contaminsate DNA solution and the inside of the quartz dish with DNase and RNase at this step.

6. Place a set of the dish onto a platform of the stand and allow it to be irradiated for an appropriate time.

Note: 5000 J/m^2: irradiation 1000 secs with UV lamp (5 J/m^2/sec)
10000 J/m^2 : irradiation 2000 secs with UV lamp (5 J/m^2/sec)
12000 J/m^2 : irradiation 2400 secs with UV lamp (5 J/m^2/sec)

7. Take out the dish and put it on a sheet of Parafilm.
8. Using forceps, remove the cover and collect the DNA solution.
9. Immediately after use of the dishes, immerse them into sterile DDW. Don't dry up completely.(*)

Note: (*) If the quartz dish is dried up completely before rinsing or if you feel the dish was stained with DNA or unwanted objects, skip to the next step.
1*. Immerse the quartz dish into 2% Adosolve in DDW for 30 mins at RT.
2*. Rinse them three times with DDW for 10 mins each.
3*. Immerse them into 0.2% DEPC for 20 mins at RT.
4*. Rinse them three times with DDW for 10 mins each.
5*. Carry out steps 4 and 5 thoroughly.

10. Rinse the quartz dishes two more times with sterile DDW.
11. Rinse them once with ethanol.
12. Return them to the original plastic dish and allow to dry.

Immunohistochemical detection

1. Cut 1x4 cm strips of nitrocellulose filters and immerse in DDW for 30 mins at RT.
2. Replace DDW for 20x SSC and keep them for 30 mins at RT.
3. Place the filters on a sheet of Parafilm and dry up.
4. Apply 2 μl of DNA solution containing DNA at various quantities per dot (10 ng,1 ng,100 pg, 10 pg, 1 pg).
5. Dry up completely (overnight air-dry is recommended).
6. Bake the filter at 80°C for 2 hrs exactly in an oven.
7. Place the filter inside a vinyl bag and incubate with a blocking solution consisting of 5% BSA (98-99% pure), 500 μg/ml normal goat IgG, 100 μg/ml yeast tRNA, 100 μg/ml salmon sperm DNA, (0.05% NaN_3) and PBS, for 1hr at RT.
8. Take the filter out of bag and replace the blocking solution with the 1st antibody (mouse anti-T-T IgG; 1:80) dissolved in PBS containing 5% BSA, 100 μg/ml yeast tRNA, 100 μg/ml salmon sperm DNA and 0.05% NaN3, for 3 hrs at RT in a vinyl bag.
9. Wash the filter with 0.075% Brij 35 in PBS at RT (4 x 15 mins).
10. Rinse the filter once with PBS.
11. React the filter with the 2nd antibody (HRP-goat anti-mouse IgG) dissolved in PBS containing 5% BSA, 100 μg/ml yeast tRNA and 100 μg/ml salmon sperm DNA, for 1hr at RT.
12. Repeat steps 9 and 10
13. Immerse the filter in the DAB/H_2O_2/Co/Ni solution for 10 mins at RT (Adams 1981).
14. Wash the filter at least 3 times with DDW for 10 mins each and dry up.

Dot blot hybridization

Day 1

1. Prepare strips of nitrocellulose filters.
2. Denature DNA by boiling for 7 mins.

Note: Use of DNA dissolved in DDW alone or excessively prolonged incubation will result in fragmentation of the DNA.

3. Apply 2 μl of DNA solution containing DNA at various quantities (10 ng,1 ng,100 pg, 10 pg, 1 pg).

Note: Be sure not to use the T-T dimered cDNA at this time.

4. Dry the filter completely (overnight air-dry is recommended).

Day 2

5. Bake the filter at 80°C for 2 hrs.
6. **Prehybridization.** Prehybridize the filter for 2 hrs at 42°C inside a vinyl bag (250 μl)(see Table 1. Prehybridization solution).

Table 1. Prehybridization solution

Hybridization mediums	Reagents	volume (μl)
10mM Tris/HCl (pH 7.4)	1M Tris/HCl (pH 7.4)	2.5
0.6M NaCl	5M NaCl	30
1mM EDTA	200mM EDTA	1.25
40% Deionized formamide	Deionized formamide	100
1x Denhardt's solution	100x Denhardt's solution	2.5
500μg/ml yeast tRNA	10mg/ml yeast tRNA	12.5
250μg/ml salmon sperm DNA	10mg/ml salmon sperm DNA	6.25
	Sterile DDW	95
	total (for 1 bag)	250

Note. The total mixture should be boiled for 7 mins.
Note. Use 250 μl for 1 filter in one vinyl bag.

7. **Hybridization.** Replace the prehybridization buffer and incubate with hybridization buffer for overnight at 42°C in the vinyl bag (see Table 2. Hybridization solution).

Table 2. Hybridization solution

Hybridization mediums	Reagents	volume (µl)
10mM Tris/HCl (pH 7.4)	1M Tris/HCl (pH 7.4)	2.5
0.6M NaCl	5M NaCl	30
1mM EDTA	200mM EDTA	1.25
40% Deionized formamide	Deionized formamide	100
1x Denhardt's solution	100x Denhardt's solution	2.5
250 µg/ml yeast tRNA	10mg/ml yeast tRNA	6.25
125 µg/ml salmon sperm DNA	10mg/ml salmon sperm DNA	3.125
10% Dextran sulfate	50% Dextran sulfate	50.0
1 µg/ml T-T dimerized probe	50µg/ml T-T dimerized probe	5.0
	TE buffer	49.375
	total (for 1 bag)	250

Note. The total mixture should be boiled for 7 mins.
Note. It may be desirable to limit Hybridization time to 12 hours.

8. Wash the filter successively as follows. **Day 3**
9. 2x SSC 15 mins, twice, RT.
10. 1x SSC 15 mins, twice, RT.
11. 0.5 x SSC 15 mins, once, RT
12. PBS, 15 mins, once, RT
13. Incubate the filter with a blocking solution consisting of 5% BSA (98-99% pure), 500 µg/ml normal goat IgG, 100 µg/ml yeast tRNA, 100 µg/ml salmon sperm DNA, 0.05% NaN_3 and PBS, for 1hr at RT in a vinyl bag.
14. Apply the 1st antibody (mouse anti-T-T IgG: 1:80) dissolved in PBS consisting 5% BSA, 100 µg/ml yeast tRNA, 100 µg/ml

salmon sperm DNA and 0.05% NaN_3 for 3 hrs at RT in vinyl bag.

15. Wash the reacted filter 4 times in PBS for 15 mins each at RT.
16. Apply the 2nd antibody (HRP-goat anti mouse IgG) dissolved in PBS consisting 5% BSA, 100 μg/ml yeast tRNA and 100 μg/ml salmon sperm DNA for 1 hr at RT in a vinyl bag.
17. Repeat step 11.
18. Immerse the filter in the DAB/H_2O_2/Co/Ni solution for 10 mins at RT (Adams 1981).
19. Wash the filter at least 3 times with DDW for 10 mins each and allow it to dry.

Localization In Situ of Specific mRNA Using T-T Dimerized Olio-DNA Probes

Recommended protocol

Note: The contamination with DNase and RNase should be minimized throughout the in situ hybridization procedure. The mRNAs are very unstable, so it is necessary to inhibit endogenous RNase activity as much as possible and to remove RNase contaminsation of the specimens and all equipment. All glassware is baked at 240°C for at least 4 hrs, and the solutions (PBS and DDW) and disposable tips are autoclaved at 121°C for 20 mins. The wearing of disposable plastic gloves (not steriled) is recommended throughout the experiment. Take care when handling H_2O_2 and DAB solutions because they are thought to be carcinogen.

Note: To estimate the adequate amount of mRNA level in paraffin sections, Methyl Green/Pyronin Y staining (Shulte 1992) should be used before performing in situ hybridization in every run.

Unless otherwise specified, all steps should be performed at room temperature.

Pretreatment

1. Cut sections to a thickness of 5 μm and place them on silane-coated glass slides.

2. Paraffin embedded sections: warm at 60°C (at least 30 mins.), deparaffinize through toluene (4x 5 mins) and then rehydrate with graded ethanol solutions (5 mins, 100%, 100%, 95%, 90%, 80%, 70%).
3. Rinse with PBS (3x 5 mins).
4. Immerse in 0.3% H_2O_2 in methanol for 15 mins to inhibit endogenous Streptavidin-binding activity and quenching peroxidase activity, if necessary.
5. Rinse in PBS (5 mins) and wash with DDW.
6. Immerse in 0.2N HCl for 20 mins and wash successively with DDW and PBS.
7. Immerse in 0.2% Triton X-100 in PBS for 10 mins and wash with PBS.

Note: Triton X-100 enhances the staining intensity of target mRNA.

8. Treat with proteinase K (1-100 µg/ml in PBS) at 37°C for 15 mins and wash with PBS (3x 5 mins).

Note: The effect of proteinase K digestion varies with the tissue and type of fixation, so the concentration of proteinase K should be optimized by individual experiment (To optimize the concentration of proteinase K, we use 28S rRNA probe to determinse the concentration of proteinase K.). Generally, in fixed frozen sections, maximal staining is obtained at proteinase K 1µg/ml, and in paraffin-embedded sections, maximal staining is obtained at proteinase K between 10 and 100 µg/ml concentrations.

Note: Before this procedure, proteinase K medium is incubated in a water bath at 37°C for 30 mins.

9. Post-fix the slides with 4% PFA /PBS for 5 mins and wash with PBS.
10. Immerse in 2 mg/ml glycine/PBS for 15 mins twice and wash with DDW.

Note: To stop the reaction of 4% PFA/PBS, the slides should be completely immersed in this medium.

11. Incubate with 40% deionized formamide/4x SSC (over 30 mins) until used for hybridization.

Note: Deionized formamide (formamide 50 ml + Regin 2.5g): Deionization should be immediately carried out for 30 mins.

Hybridization

12. Apply 30-40 μl of the hybridization mixture on each slide and keep it in a moist chamber humidified with 40% formamide/4x SSC at 37 - 42°C for 15-17 hrs.

Note: Before hybridization, the mixture is heat-denatured for 7 mins.
Note: The moist chamber should be completely shut off by masking tape.

Stringent washing

13. Wash with 50% formamide/2x SSC/0.075% Brij 35 (3 x 1 hr) at 37°C and with 2x SSC (2 x 15 mins).

Note: For a more stringent washing 0.5 x SSC and 0.2 x SSC should be used.

Enzyme immunohistochemistry

14. Apply 30-40 μl of the blocking solution and leave for 1 hr in a moist chamber humidified with PBS.

Note: To minimize the extent of contaminsation with DNase and RNase, care should be taken during handling of the specimens.

15. Apply 30-40 μl of the blocking solution containing HRP-labeled anti-(T-T dimer) monoclonal antibody (1:80) for 15-17 hrs in a moist chamber.
16. Wash 4 times with 0.075% Brij 35/PBS for 15 mins each.
17. After washing with PBS, apply chromogen solution (DAB, H_2O_2, nickel and cobalt) to the specimens and monitor the reaction under a light microscope for optimum signal and stop the reaction immediately by immersing in water once the satisfactory color develops (Adams 1981).
18. Dehydrate with serial up-graded ethanol solutions, treat with xylene and mount without any counterstaining.

Results

Dot blot hybridization with T-T dimerized probes

Dot blot hybridization using T-T dimerized sense and antisense KGFR oligo-DNA probes was shown in Fig 1. Aliquots (2μl) of KGFR sense oligo-DNA solution were dotted on nitrocellulose filters at 1 pg to 10 ng /spot. When T-T dimerized KGFR antisense oligo-DNA probe was hybridized with KGFR sense oligo-DNA on a nitrocellulose filter, at least 10 pg of the oligo-DNA was detected, while no signal was seen with dotted KGFR sense oligo-DNA.

In situ hybridization with T-T dimerized probes

In situ hybridization using T-T dimerized Metallothionein oligo-DNA probes and 28S rRNA complementary oligo-DNA probe in paraffin sections of esophageal cancer are shown in Figure 2.

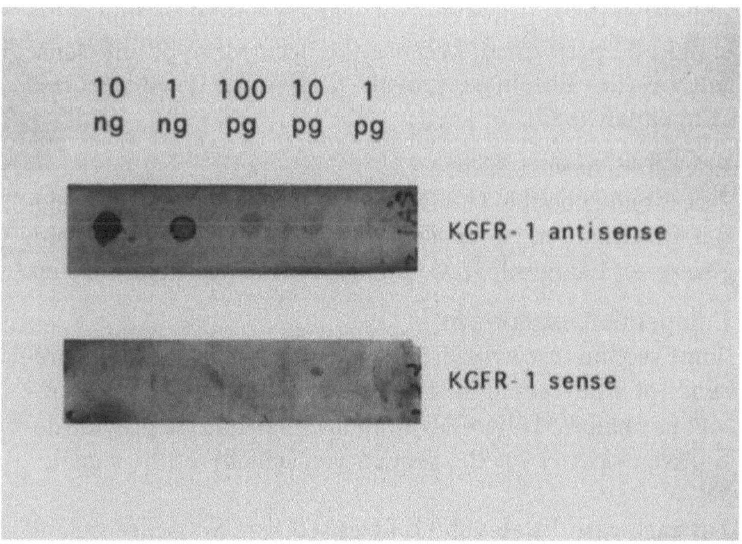

Fig. 1. Colorimetric dot blot hybridization. Various amouts (1 pg to10 ng/spot) of KGFR sense oligo-DNA, which were fixed onto nitrocellulose membranes, were hybridized with T-T labeled KGFR antisense oligo-DNA or T-T labeled KGFRsense oligo-DNA. At least 10 pg of the antisense oligo-DNA was detected.

To verify the results obtained by in situ hybridization, various control experiments are required.

1. Evaluation of levels of hybridizable RNAs
   To evaluate the levels of hybridizable RNAs in tissue sections, we used 28S rRNA probe in every run. This oligo-DNA is a sequence complementary to the 28S rRNA sequence (nucleotide nos. 2173-2206) of rat rRNA, which are found to have 100% homology with mouse, rat and human. The oligo-DNA added with 3 and 2 repeats of adenine-thymine-thymine (ATT) at the 5' and 3' ends, respectively for the thymine-thymine (T-T) dimers. In situ localization of hybridizable rRNA for the positive control was described previously in detail (Yoshii 1995).

2. Positive control
   As a positive control, tissues and/or cells which were already reported to express target mRNA should be used.

3. Negative control
   Sense probe should be used as a negative control in every run. Although when using oligo-DNA probes the sense strand is utilized as the negative control probe, a careful examinsation should be performed, because the occurrence of anti-sense mRNA (i.e. fibroblast growth factor) has been reported (Kimelman 1989).

4. Pretreatment with RNase
   To estimate possible involvement of proteins and DNA in signal information, some sections should be digested with RNase-A (100µg/ml, 37°C, 1hr) before the post-fixation step.

5. Competition experiment
   Some sections are hybridized with antisense probe in the presence of an excess amount of either homologous or heterologous unlabeled oligo-DNA (about 10-20 fold) to provide definitive evidence for the sequence specificity of the signal.

Comments

- For each case 2 slides should be used, one for antisense and one for sense probe to compare the staining intensity between the sense and antisense probes. Another should be used for 28S rRNA probe which estimates the amount of hybridizable RNA level in every run.

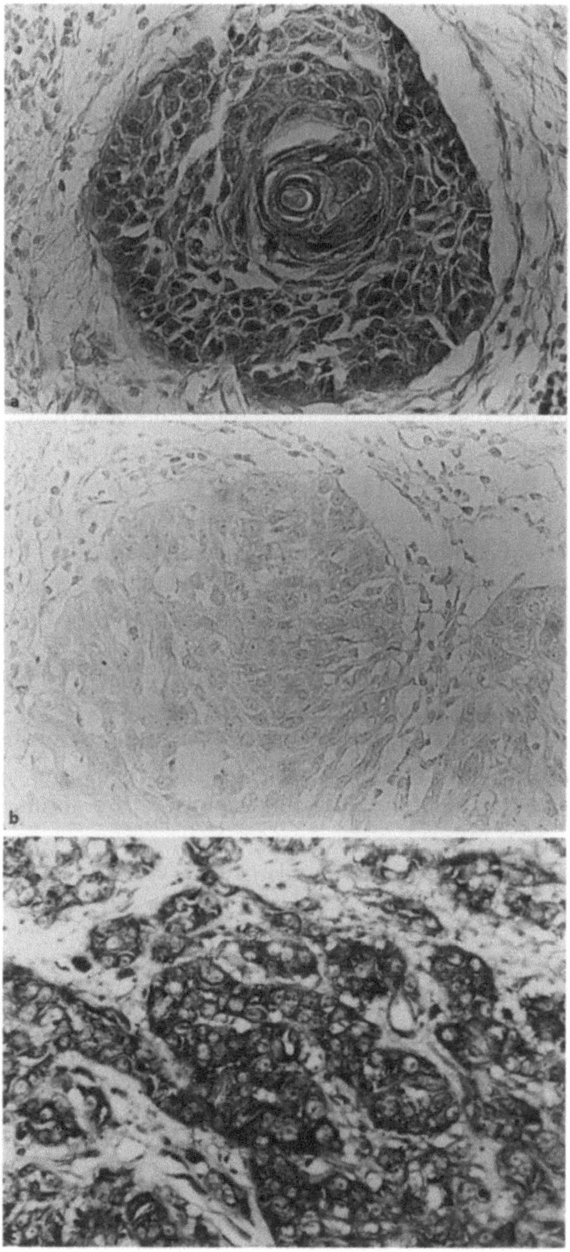

Fig. 2. In situ hybridization using Metallothionein (MT) antisense (a), MT sense (b), and 28S rRNA antisense (c) oligo-DNA probes in paraffin sections of the esophageal cancer. (a): positive staining was found in the cancer. (b): no positive staining was detected in the same section. (c): Adequate amounts of hybridizable RNAs were detected in paraffin section.

- To determine the optimal condition of in situ hybridization, positive control probe should be used.

Troubleshooting

- With the antisense probe, none or weak staining is found.
  - Inadequate amount of mRNA levels in the tissues : Check the staining of Methyl Green/Pyronin Y and 28S rRNA probe as a positive control.
  - The concentration of the antisense probe : Raise the concentration of the antisense probe.
  - The concentration of proteinase K : Change the concentration of proteinase K (e.g. a lower or higher concentration).
  - The optimal concentration of the antibody : Carry out the titration of the antibody using immunodetection method to verify whether the probes are optimally labeled.
  - If you use direct enzyme immunohistochemistry as visualization system, you may change it to indirect method.
- With the sense and/or antisense probes, background staining is found.
  - Stringent washing : Use a more stringent washing at lower concentrations of salt (e.g. 0.5 x SSC, 0.2 x SSC).
  - Hybridization temperature : Raise the temperature about 5°C.
  - Hybridization mixtures : Add heparin and SDS to the mixture.

References

Adams JC (1981) Heavy metal intensification of DAB-based HRP reaction products. J Histochem Cytochem 29:775

Feavers IM, McEwan IJ, Liang H, Jost J-P (1989) An estradiol-dependent protein from chicken liver binds single-stranded DNA and RNA. J Bio Chem 264:9114-9117

Franklin WA, Lo KM, Haseltine WA (1982) J Biol Chem 257:13535-13543

Kimelman D, Kirschner MW (1989) An antisense mRNA directs the covalent modification of the transcript encoding fibroblast growth factor in Xenopus oocytes. Cell 59:687-696

Koji T, Moriuchi T, Nakane PK (1988) Improved tissue preparation for in situ localization of specific mRNA using non-radioactive DNA probes : Effects of protease digestion and probe size on signal detection in frozen and paraffin sections of rat pituitary glands. Acta Histochem Cytochem 21:187-200

Koji T, Nakane PK (1990) Localization in situ of specific mRNA using Thymine-Thymine dimerized DNA probes: sensitive and reliable non-radioactive in situ hybridization. Acta Pathol Jpn 40:793-807

Koji T, Nakane PK (1996) Recent advances in molecular-histochemical-techniques: in situ hybridization and southwestern histochemistry. J Electron Microsc 45:119-127

Koji T, Sugawara I, Kimura M et al. (1989) In situ localization of c-myc mRNA in HL-60 cells using non radioactive synthetic oligonucleotide probes. Acta Histochem Cytochem 22:295-307

Meinkoth J, Wahl G (1984) Hybridization of nucleic acids immobilized on solid supports. Anal Biochem 138:267-284

Shulte EKW, Lyon HO, Hoyer PE (1992) Simultaneous quantification of DNA and RNA in tissue sections. A comparative analysis of the methyl green-pyronin technique with the gallocyanin chromalum and Feulgen procedures using image cytometry. Histochem J 24:305-310

Singh L, Jones KW (1984) The use of heparin as a simple costeffective means of controlling background in nucleic acid hybridization procedures. Nucleic Acids Res 12:5627-5638

Yoshii A, Koji T, Ohsawa N, Nakane PK (1995) In situ localization of ribosomal RNAs is a reliable reference for hybridizable RNA in tissue sections. J Histochem Cytochem 43:321-328

In Situ Hybridization for RNA: Nonradioactive Probe: Oligo-DNA Probe: Digoxigenin (1)

YUKA IWASAKI AND IHACHI KATO

Introduction

In this article, protocols are described for preparation of digoxigenin-labeled oligo-DNA probe and their use in in situ hybridization. As has been shown previously (Koji et al. 1988, 1989; Koji and Brenner 1993; Koji and Nakane 1990, 1996), nonradioactive synthetic oligo-DNA probes can be very useful tools. The 3' -end of oligo-DNA can be labeled enzymatically with several possible haptens, such as digoxigenin and biotin. Digoxigenin, a plant steroidal aglycone, is especially suitable for immunocytochemical detection because its epitope is not present in animal cells. Also, it has recently been confirmed that under appropriate conditions, there is no substantial difference in the sensitivity of detection between [^{35}S] cRNA (Denjin et al. 1990) or [^{35}S] cDNA (Unger et al. 1991) probes and nonradioactive probes.

In nonradioactive in situ hybridization, notable improvements have been made in the protocol, including the visualization systems, resulting in greater detection sensitivity. The attempts to quantify colorimetric signals by computer-assisted image analysis are also progressing. (The result shown in this chapter is obtained by the use of an image analyzer.)

Materials

for Preparation of tissues

- paraformaldehyde (PFA) (MERCK)

Correspondence to Yuka Iwasaki, Nagasaki University School of Dentistry, Department of Periodontology, 1-7-1 Sakamoto, Nagasaki, 852, Japan (*phone* +81-95-849-7683; *fax* +81-95-849-7684)

for Oligo-DNA tailing

- oligo-DNA probe
- digoxigenin (DIG) oligonucleotide tailing kit (Boehringer Mannheim)

for Immuno-histochemical detection of probes on nitrocellulose membranes

- UV lamp (254 nm wave length) (Toshiba Ltd., Tokyo, Japan)
- UV radiometer (taking an accurate measurement of 5 $J/m^2/sec$)(Tokyo, Kougaku Kikai, Ltd., Tokyo, Japan)
- nitrocellulose filter
- vinyl bag
- 20 x SSC (1 x SSC = 0.15 M NaCl, 0.015 M sodium citrate, pH 7.0)
- bovine serum albumin (BSA, 98-99% pure) (Sigma)
- normal sheep IgG (Sigma)
- yeast tRNA (Sigma)
- salmon testis DNA (Sigma)
- horseradish peroxidase (HRP)-labeled sheep anti-digoxigenin antibody (Boehringer Mannheim)
- The blocking solution containing 500 μg/ml normal sheep IgG, 5% BSA, 100 μg/ml yeast tRNA, 100 μg/ml salmon testis DNA in PBS.
- The chromogen solution containing 3,3' -diaminobenzidine (DAB), H_2O_2, nickel and cobalt; 0.5 mg/ml DAB/0.025% $CoCl_2$/0.02% $NiSO_4$ $(NH_4)_2SO_4$/0.01% H_2O_2 /0.1 M phosphate buffer (pH 7.5). (Adams 1981)

for Dot blot hybridization

- UV lamp (254 nm wave length)
- UV radiometer (taking an accurate measurement of 5 $J/m^2/sec$)
- nitrocellulose filter
- vinyl bag
- 20x SSC
- formamide (nacalai tesque)
- Mixed bed resin [AG-501-X8 (D) ; Bio-Rad]
- Denhardt's solution [0.02% BSA (98-99% pure), 0.02% Ficoll-400 (MW=400000) (Pharmacia Fine Chemicals), 0.02% polyvinylpyrolidone (MW=360000) (Sigma)]
- dextran sulfate (MW=500000) (Sigma)
- yeast tRNA
- salmon testis DNA
- BSA
- HRP-labeled sheep anti-digoxigenin antibody

- The constituents of prehybridization mixture; 10 mM Tris/ HCl buffer (pH 7.4), 0.6 M NaCl, 1 mM EDTA, 1 x Denhardt's solution, 500 μg/ml yeast tRNA, 250 μg/ml salmon testis DNA, 40% deionized formamide[a)] and 1 μg/ml digoxigenin-labeled oligo-DNA probe. It is heat-denatured for 7 mins before hybridization.
  a) Formamide is deionized with mixed bed resin (formamide 50 ml + mixed bed resin 2.5 g).
- The constituents of hybridization mixture; 10 mM Tris/ HCl buffer (pH 7.4), 0.6 M NaCl, 1 mM EDTA, 1 x Denhardt's solution, 250 μg/ml yeast tRNA, 125 μg/ml salmon testis DNA, 10% dextran sulfate, 40% deionized formamide[a)] and 0.5-2 μg/ml digoxigenin-labeled oligo-DNA probe. It is heat-denatured for 7 mins before hybridization.
  a) Formamide is deionized with mixed bed resin.
- The blocking solution containing 500 μg/ml normal sheep IgG, 5% BSA, 100 μg/ml tRNA, 100 μg/ml salmon testis DNA in PBS
- The chromogen solution containing 3,3' -diaminobenzidine (DAB), H_2O_2, nickel and cobalt; 0.5 mg/ml DAB/0.025% $CoCl_2$/0.02% $NiSO_4$ $(NH_4)_2SO_4$/0.01% H_2O_2 /0.1 M phosphate buffer (pH 7.5)

for In situ hybridization

- PFA
- proteinase K (Wako)
- formamide (nacalai tesque)
- mixed bed resin [AG-501-X8 (D) ; Bio-Rad]
- Denhardt's solution [0.02% BSA (98-99% pure), 0.02% Ficoll-400 (MW=400000), 0.02% polyvinylpyrolidone (MW = 360000)]
- yeast tRNA
- salmon testis DNA
- dextran sulfate (MW=500000)
- 20 x SSC
- normal sheep IgG
- HRP-labeled sheep anti-digoxigenin antibody
- The constituents of hybridization mixture; 10 mM Tris/ HCl buffer (pH 7.4), 0.6 M NaCl, 1 mM EDTA, 1 x Denhardt's solution, 250 μg/ml yeast tRNA, 125 μg/ml salmon testis DNA, 10% dextran sulfate, 40% deionized formamide and 0.5-2 μg/ml digoxigenin-labeled oligo-DNA probe. It is heat-dena-

tured for 7 mins before hybridization. In some cases, probes are used as cocktails consisting of equimolar mixtures.
a) Formamide is deionized with mixed bed resin.

- The blocking solution containing 500 μg/ml normal sheep IgG, 5% BSA, 0.3 M NaCl, 100 μg/ml tRNA, 100 μg/ml salmon testis DNA in PBS
- The chromogen solution containing 3,3' -diaminobenzidine (DAB), H_2O_2, nickel and cobalt; 0.5 mg/ml DAB/0.025% $CoCl_2$/0.02% $NiSO_4$ $(NH_4)_2SO_4$/0.01% H_2O_2 /0.1 M phosphate buffer (pH 7.5)

Procedure

Unless otherwise specified, all steps are performed at room temperature.

Preparation of tissues

The tissues are fixed in 4% PFA/PBS, and embedded in paraffin according to a routine procedure. The specimens containing bone or teeth have to be decalcified before being processed to usual histochemical techniques. The influence of decalcification procedure on the retention of hybridizable RNA is an important consideration. Therefore, to achieve an optimal balance between retention of hybridizable 28S rRNA and morphological preservation, we compared a variety of decalcification protocols in Table 1. These results indicate that the decalcification protocol with 10% EDTA at 4°C for 5-6 days is the best.

Table 1. Various decalcification protocols

Decalcification	Signal intensity	Morphology
10 % EDTA (4 °C, 5-6 days)	+++	excellent
5 % Formic acid (4 °C, 24 h)	+++	good
5 % HCl (4 °C, 24 h)	+	excellent
Plank-Rychlo solution (4 °C, 24 h)	++	excellent
10 % Nitric acid (4 °C, 24 h)	++	poor

Various decalcification protocols were compared for the signal intenstiy of 28S rRNA. Estimation was performed based on the most strongly stained cells. Rating scale of signal intensity: +, weak; ++, definite; and +++, intense.

The recommended protocol is described for preparation of the decalcified tissues.

1. Fix in 4% PFA/PBS at 4 °C for 12 h and wash with DDW.
2. Decalcify with 10% EDTA at 4°C for 5-6 days and wash with DDW at 4°C for 12 h.
3. Dehydrate with graded ethanols and xylene, and embed in paraffin wax.

Synthesis and labeling of oligo-DNAs

Oligo-DNA probes require about 45-base sequence complementary to the target mRNAs. We chose a sequence according to the published cDNA base sequence data and synthesized it on a DNA synthesizer, or obtained it commercially. We also prepared an oligo-DNA with 34 bases, which is complementary to the part of 28S rRNA, as a positive control probe, as described in detail by Yoshii et al. (1995). This control probe is used to assess the integrity and hybridizability of RNA in tissue sections. In addition, we conducted a computer-assisted search of the oligo-DNA sequences and no significant homology with any other published sequence was found.

The oligo-DNAs are labeled by a DIG oligonucleotide tailing kit according to standard protocols. After separation from the unreacted compounds by ethanol precipitation, the final precipitates are suspended in TE buffer (10 mM Tris-HCl, pH 7.4, containing 1 mM EDTA) at a final concentration of 25 μg/ml.

Immunohistochemical detection of probes on nitrocellulose membranes

To confirm the labeling of probe oligo-DNAs, we perform an immunohistochemical detection method.

1. Cut 1 x 4 cm strips of nitrocellulose filters and immerse in DDW for 30 mins at RT.
2. Replace DDW for 20 x SSC and keep them for 30 mins at RT.
3. Place the filters on a sheet of Parafilm and dry up.
4. Apply 2 μl of DNA solution containing digoxigenin-labeled oligo-DNAs onto the filter at various quantities per dot (1 pg, 10 pg, 100 pg, 1 ng, 10 ng).
5. Dry up completely (overnight air-dry is recommended).
6. Bake the filter at 80°C for 2 hrs.
7. Irradiate the filter for 1 min.

Note: The appropriate dose of irradiation is 300-600 J/m^2. 5000 J/m^2 : irradiation 1000 secs with UV lamp (5 J /m^2/sec). 10000 J/m^2 : irradiation 2000 secs with UV lamp (5 J /m^2/sec)

8. React the filter with a blocking solution for 1 hr at RT in a vinyl bag. (250 μl/1-2 filters)
9. React the filter with the blocking solution including HRP-labeled sheep anti-digoxigenin antibody (1: 400-800) for 3 hrs in a vinyl bag. (250 μl/1-2 filters)
10. Wash 4 times with 0.075% Brij 35/PBS for 15 mins each at RT.
11. After a brief washing with PBS, visualize the HRP sites in a chromogen solution containing DAB, H_2O_2, nickel and cobalt at RT.
12. Wash the filter at least 3 times with DDW for 5 mins each and dry up.

Dot blot hybridization

To confirm the specificity and the adequate sensitivity of probe oligo-DNAs, dot blot hybridization should be performed before being processed to in situ hybridization study.

1. Prepare strips of nitrocellulose filters according to the steps 1-3 of the instruction for immunohistochemical detection.
2. Apply 2 μl of DNA solution containing unlabeled oligo-DNAs onto the filter at various quantities per dot (1 pg, 10 pg, 100 pg, 1 ng, 10 ng).
3. Dry up completely (overnight air-dry is recommended).
4. Bake the filter at 80°C for 2 hrs.
5. Irradiate the filter for 1 min.
6. Prehybridize the filter with a prehybridization mixture for 2 hrs at 37-42°C in a vinyl bag . (250 μl/1-2 filters)
7. Hybridize the filter with a hybridization mixture for 15-17 hrs at 37-42°C in the vinyl bag . (250 μl/1-2 filters)
8. Wash the filter successively as follows; 2 x SSC (15 mins, twice, RT), 1 x SSC (15 mins, twice, RT), 0.5 x SSC (15 mins, once, RT) and PBS (15 mins, once, RT).

9. React the filter with a blocking solution for 1 hr at RT in a vinyl bag. (250 µl/1-2 filters)

10. React the filter with the blocking solution including HRP-labeled sheep anti-digoxigenin antibody (1: 400-800) for 3 hrs in a vinyl bag. (250 µl/1-2 filters)

11. Wash 4 times with 0.075% Brij 35/PBS for 15 mins each at RT.

12. After a brief washing with PBS, visualize the HRP sites in a chromogen solution containing DAB, H_2O_2, nickel and cobalt at RT.

13. Wash the filter at least 3 times with DDW for 5 mins each and dry up.

In situ hybridization

We confirm the effectiveness of 28S rRNA staining as a reference for optimization of the pretreatment conditions including protease digestion.

1. Cut sections to a thickness of 5-6 µm and place them on silane-coated glass slides according to routine procedure.

2. Prior to use, warm at 60°C for at least 30 mins, deparaffinize through toluene and then rehydrate with serial ethanol solutions and wash 3 times with PBS for 5 mins each.

3. Immerse in 0.3-3% H_2O_2 in methanol for 15 mins to inhibit endogenous peroxidase activity, if necessary.

4. After a brief washing with PBS and DDW, immerse in 0.2 N HCl for 20 mins and wash successively with DDW and PBS.

5. Treat with proteinase K (1-100 µg/ml in PBS) at 37°C for 10 mins and wash 3 times with PBS.

Note: Proteinase K solution is incubated in a water bath at 37°C for 30 mins before this procedure. It is stored at -20°C in small aliquots, which is used once and discarded. The effect of proteinase K digestion varies with the tissue and type of fixation, so the concentration of proteinase K should be optimized with individual experiment. In addition, the optimal concentration of proteinase K in the decalcified tissue sections (5-10 µg/ml) is clearly lower than that for paraffin-embedded sections of noncalcified tissue, i.e., 50-100 µg/ml.

6. Fix in 4% PFA/PBS for 5 mins and wash twice with PBS.
7. Immerse twice in 2 mg/ml glycine/PBS for 15 mins each.
8. After brief washings with PBS and DDW, successively, immerse in 40% deionized formamide/4 x SSC until used for hybridization.
9. Put 20-30 µl of the hybridization mixture on each slide and keep it in a moist chamber humidified with 40% formamide/4 x SSC at 37°C for 15-17 hrs. Under these conditions, no covers to avoid evaporation of the medium are needed.

Note: Furthermore, the addition of 200 U/ml heparin and/or 0.01% SDS to the hybridization mixture often decreased the nonspecific reaction of the probes.

10. Wash 5 times with 0.075% Brij 35/2 x SSC for 1 hr each time at 37 °C and twice with 2 x SSC for 15 mins each.

Note: If necessary, 50% formamide is included, and for a more stringent washing lower concentrations of salt such as 0.5 x SSC and 0.2 x SSC are used.

11. Immerse in 70% ethanol (5 mins), 90% ethanol (5 mins), 100% ethanol (5 mins, twice), and acetone (5 mins, twice), successively, as a delipidation.

Note: This step after hybridization is effective in removing the probe bound to sections nonspecifically (Koji et al. 1994), especially in some cases including decalcified tissues (Iwasaki et al.1998).

12. After a brief washing with PBS, put 30-40 µl of a blocking solution and leave to stand for 1 hr in a moist chamber humidified with PBS.
13. React with 30-40 µl of the blocking solution including HRP-labeled sheep anti-digoxigenin antibody (1: 100-200) for 15-17 hrs in the moist chamber.
14. Wash 4 times with 0.075% Brij 35/PBS for 15 mins each.
15. After a brief washing with PBS, visualize the HRP sites in a chromogen solution containing DAB, H_2O_2, nickel and cobalt.
16. Dehydrate with serial up-graded ethanol solutions, cleared with xylene and mounted without any counterstaining.

Results

To confirm the specificity of mRNA signals obtained by in situ hybridization, a variety of control experiments are required. In every run, sense and 28S rRNA probe are used as negative and positive control, respectively. To exclude possible involvement of proteins and DNA in signal formation, some sections are digested with RNase A (100 µg/ml) at 37°C for 1 hr, before hybridization. Others are hybridized with antisense probe in the presence of an excess amount of unlabeled antisense or sense oligo-DNA, while some are washed under extremely high stringent conditions after hybridization, to provide definitive evidence for the sequence specificity of the signals.

Fig. 1 shows an example of detection of 28S rRNA and IL-10 mRNA by this protocol using digoxigenin-labeled oligo-DNA probes. The sections are paraffin embedded after fixation with 4% PFA/PBS and decalcification with 10% EDTA. The cells positive for each mRNA are evaluated by the measurement of the staining density more than the level of staining by negative con-

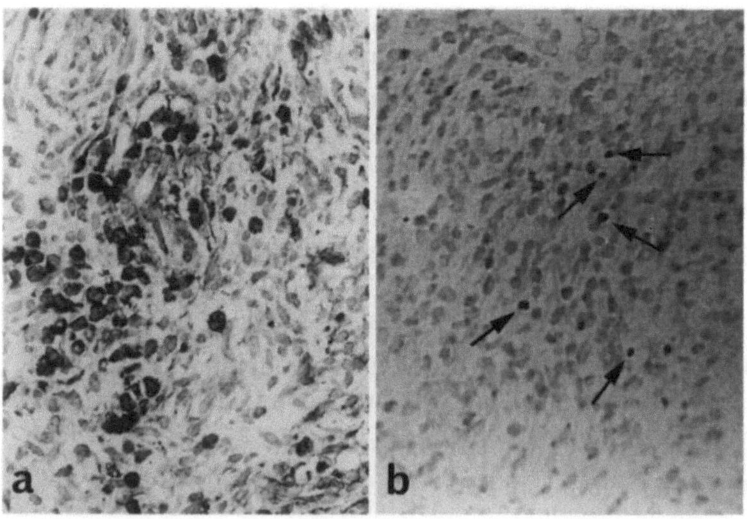

Fig. 1. Distribution of cells positive for 28S rRNA and IL-10 mRNA in gingival tissue of mouse lipopolysaccharide-induced periodontitis mandibles, assessed by in situ hybridization. (a) A section is hybridized with 28S rRNA complementary probe. (b) A section is hybridized with IL-10 antisense probe. Arrows; positive cells. (Magnification x 350)

trol probe using an image analyzer (IBAS 2000 system, Carl Zeiss).

References

Adams JC (1981) Heavy metal intensification of DAB-based HRP reaction product. J Histochem Cytochem 29: 775

Denjin M, De Weger RA, Berends MJH, Compier-Spies PI, Jansz H, Van Unnik JAM, Lips CJM (1990) Detection of calcitonin-encoding mRNA by radioactive and non-radioactive *in situ* hybridization: improved colorimetric detection and cellular localization of mRNA in thyroid sections. J Histochem Cytochem 38: 351-358

Iwasaki Y, Hara Y, Koji T, Shibata Y, Nakane PK, Kato I (1998) Differential expression of IFN-γ,IL-4, IL-10, and IL-1β mRNAs in decalcified tissue sections of mouse lipopolysaccharide-induced periodontitis mandibles assessed by in situ hybridization. Histochem and Cell Biol 1090:339-347

Koji T, Brenner RM (1993) Localization of estrogen receptor messenger ribonucleic acid in rhesus monkey uterus by nonradioactive in situ hybridization with digoxigenin-labeled oligodeoxynucleotides. Endocrinology 132: 382-392

Koji T, Nakane PK (1990) Localization *in situ* of specific mRNA using thymine-thymine dimmerized DNA probes. Sensitive and reliable non-radioactive *in situ* hybridization. Acta Pathol Jpn 40: 793-807

Koji T, Nakane PK (1996) Recent advances in molecular histochemical techniques: in situ hybridization and southwestern histochemistry. J Electron Microsc 45: 119-127

Koji T, Chedid M, Rubin JS, Slayden OD, Csaky KG, Aaronson SA, Brenner RM (1994) Progesterone-dependent expression of keratinocyte growth factor mRNA in stromal cells of the primate endometrium: keratinocyte growth factor as a progestomedin. J Cell Biol 125: 393-401

Koji T, Moriuchi T, Nakane PK (1988) Improved tissue preparation for *in situ* localization of specific mRNA using non-radioactive DNA probes: effects of protease digestion and probe size on signal detection in frozen and paraffin sections of rat pituitary glands. Acta Histochem Cytochem 21: 187-200

Koji T, Sugawara I, Kimura M, Nakane PK (1989) *In situ* localization of *c-myc* mRNA in HL-60 cells using non-radioactive synthetic oligodeoxynucleotide probes. Acta Histochem Cytochem 22: 295-307

Unger ER, Hammer ML, Chenggis ML (1991) Comparison of ^{35}S and biotin as labels for *in situ* hybridization: use of an HPV model system. J Histochem Cytochem 39: 145-150

Yoshii A, Koji T, Ohsawa N, Nakane PK (1995) In situ localization of ribosomal RNAs is a reliable reference for hybridizable RNA in tissue sections. J Histochem Cytochem 43: 321-327

In Situ Hybridization for RNA: Nonradioactive Probe: Oligo-DNA Probe: Digoxigenin (II)

MASANOBU MIYAZAKI, YOSHIYUKI OZONO, TAKASHI HARADA AND SHIGERU KOHNO

Introduction

Nonradioactive in situ hybridization is thought to be less sensitive than the radioactive method, especially when used with oligo-DNA probe. However, the sensitivity of nonradioactive in situ hybridization using oligo-DNA is sufficient for localization of mRNA at cellular level even in human biopsy specimens. The method of nonradioactive in situ hybridization was developed in our laboratory in 1994 (Miyazaki et al. 1994) and was applied to identify and localize mRNA of several factors synthesized by renal cells in normal individuals and patients with a variety of renal diseases (Suzuki et al. 1995; Miyazaki et al. 1996). Nonradioactive in situ hybridization with oligo-DNA has several advantages compared with the radioactive method, including a high resolution, good probe stability and no exposure to radiation, as described in detail by Koji and Nakane (1996). Compared with other types of probes, such as RNA or cDNA probe, oligo-DNA has several benefits. First, oligo-DNA is easier to prepare. While several years ago, a special DNA synthesizer that was available only in certain laboratories, was necessary for the preparation of oligo-DNA, such synthesizers are less expensive nowadays. Alternatively, custom made probes can be purchased nowadays from several commercial sources. Second, a special knowledge of molecular technology and expertise, e.g., cloning techniques, is not necessary. Finally, the probes can be easily de-

Correspondence to Masanobu Miyazaki, Nagasaki University School of Medicine, The Second Department of Internal Medicine, 1-7-1 Sakamoto, Nagasaki, 852-8501, Japan (*phone* +81-95-849-7273; *fax* +81-95-849-7285; *e-mail* msnbmiya-ngs@umin.ac.jp)

signed. The experimenter can also select the sequence, which is well conserved among different species. Thus, the selected sequence allows the use of one probe on different sections from different species.

In nonradioactive in situ hybridization, the probe is labeled with a hapten, such as biotin, digoxigenin (DIG) and thymine-thymine (T-T) dimers. The T-T dimers method was originally described by Drs. Koji and Nakane (1990). The in situ hybridization method using T-T dimerized oligo-probe is described in → Chapter 11. Biotin labeled probe is used in nonradioactive in situ hybridization, however, endogenous biotin, also called vitamin H, is present in large quantities in certain organs, including kidney, liver and muscle. Since biotin is inactivated by aldehyde, the presence of biotin in these tissues does not cause a problem when tissues are fixed with aldehyde and embedded in paraffin. While fresh frozen tissues are used for in situ hybridization, blocking of endogenous biotin is an indispensable step in this method. Endogenous biotin blocking kit is available from Vector Company. In contrast, since DIG does not exist in animal and human body, blocking is not necessary, i.e., no background contamination is present as a result of cross-reaction with endogenous substances.

In this chapter, we describe the protocol of nonradioactive in situ hybridization using DIG labeled oligonucleotide. The described method is especially suitable for processing renal biopsy specimens.

Outline

The entire protocol of nonradioactive in situ hybridization with oligo-DNA probe is as follows:

1. Preparation of the tissue
2. Labeling probe
3. Hybridization with probe
4. Visualization of the bound probe on the section

This outline may be almost similar to other protocols of in situ hybridization method. The three most important points are as follows. First, contamination of RNase should be avoided at all steps of the first day. Second, sections must not be dried at any step once the procedure of in situ hybridization is started. If the sections are dry, a non-specific signal is obtained. Finally, to the beginner, be careful when handling a very small amount of the reagent, especially on the first day.

Materials

- O.C.T compound (Miles Inc., Elkhardt, IN)
- Diethylpyrocarbonate (Sigma)
- 3-aminopropyltriethoxysilane (Sigma)
- Paraformaldehyde (Merk 104005)
- Digoxigenin oligonucleotide tailing kit (Boehringer Mannheim 1417 231, Germany)
- Diethylpyrocarbonate-treated phosphate buffered saline (PBS)
- Proteinase K (Sigma P-4914)
- Salmon testis DNA (Sigma)
- Bovine serum albumin (BSA, Sigma)
- Ficoll (Sigma)
- Polyvinylpyrrolidone (Sigma)
- AG 501-X8 Resin (BIO-RAD 142-6425)
- Incubation chamber
- BRIJ 35 (23 Lauryl ether, Sigma)
- Monoclonal anti-digoxigenin antibody (Boehringer Mannheim 1333 062)
- Normal swine serum (DAKO)
- Horseradish peroxidase (HRP)-conjugated rabbit anti-mouse antibody (Dako P260)
- HRP-conjugated swine anti-rabbit antibody (Dako Z113)

Required solutions

- Diethylpyrocarbonate (DEPC)-treated water:
  Add 0.1% DEPC to distilled water and stir it well for at least one hour. Autoclave the solution prior to use.

Note: To obtain the signal, preservation of mRNA in sections is essential. It is necessary to minimize the activity of RNase in all steps of in situ hybridization as well as Northern blotting or polymerase chain reaction. All solutions should be prepared using

RNase-free glassware, autoclaved water and chemicals for molecular biology use, if possible. All solutions, except those containing Tris, should be treated with 0.1% DEPC followed by autoclaving.

- 3-aminopropyltriethoxysilane (APS)-coated slide glasses:
  Soak slide glasses in acetone dissolved with APS for 5 to 10 seconds. Wash them briefly with acetone and DEPC-treated water sequentially. Dry the slides overnight. Commercial APS-coated slide glasses are also available (Matsunami, Tokyo or DAKO)
- 4% paraformaldehyde in PBS:
  Add 40g paraformaldehyde to warm 800 ml of DEPC-treated water on a heat plate. Heat the solution and add a few drops of 1M NaOH. When paraformaldehyde dissolves completely, the color of the solution changes from cloudy to clear. Add 100 ml of 0.1 M PBS to the solution and leave to cool in ice. Then, check pH and adjust to pH 7.4 by 1M HCl.

Note: Paraformaldehyde can be dissolved only in warm alkaline solution.

Note: Preparation should be completed under the laboratory hood.

- 50 x Denhardt's solution:
  1% bovine serum albumin (Fraction V; Sigma), 1% Ficoll (Type 400, Pharmacia) and 1% polyvinylpyrrolidone in DEPC-treated water.
- Standard saline citrate (SSC):
  20x SSC; 3 M NaCl + 0.3 M trisodium citrate
- Formamide:
  Formamide for molecular biology is available from several chemical firms. However, if any yellow color is present, the formamide should be deionized by resin. After filtering, deionized formamide can be stored at -70°C.
- Prehybridization buffer:
  - 4x SSC
  - 50 mM sodium phosphate
  - 5x Denhardt's solution
  - 0.2 mg/ml salmon testis DNA
  - 0.2 mg/ml yeast tRNA
  - 20% deionized formamide

- Hybridization buffer:
  Final probe concentration is 0.5 - 2.0 ng/μl in the prehybridization buffer.
- Washing solution (WS):
  - WS1;2x SSC with 0.075% of BRIJ
  - WS2;0.5x SSC with 0.075% of BRIJ
- Preincubation buffer in histochemistry:
  - 20% normal swine serum
  - 5% BSA
  - 5% FCS in PBS
- Dilution buffer of second and third antibodies:
  - 4% normal swine serum
  - 1% BSA
  - 1% FCS in PBS

Procedure

Tissue preparation

The biopsy specimens are embedded in OCT compound and stored -70°C until use. Fresh frozen tissue is cut into 4 μm thick sections and adhered onto APS-coated slide glasses. Dry the slides well by using cold air of a dryer for at least 30 mins. Sections can be stored in a tight sealed box at -70°C.

Note: The sections can be generally stored for 6 months (Signal can be obtained).

Note: Tissue samples can be embedded in wax. Fixation before embedding is important to obtain a good signal. Strong fixation results in detailed morphology, whereas a low signal may be due to reduced accessibility of the probe to the target mRNA. We recommend the use of 4% paraformaldehyde for fixation.

Note: Assessment of remaining RNA in tissue: When a negative signal is obtained in in situ hybridization, it is difficult to determine whether this represents a lack of the specific mRNA, a technical error or whether mRNA is destroyed in the tissue. Methyl green-pyronin Y staining is easy to perform and useful to examine the retained RNA (Schulte et al. 1992). Although the pink color is stained by methyl green-pyronin-Y, indicating preservation of RNA in the tissue, too strong fixation may cause a false negative because of low accessibility of the probe. Therefore, a positive control of in situ hybridization is recommended. β-actin mRNA or poly(A) were used in the past as a positive control for in situ hybridization (reviewed by Koji and Nakane 1996). However, Yoshii et al. (1995) indicated that 28S ribosomal RNA (rRNA) has some advantages over the former controls. 28S rRNA is widely and abundantly distributed throughout the tissue and its presence in the tissues is fairly constant. Furthermore, in situ hybridization for 28S rRNA shows hybridizability of tissue RNA and this helps us to assess usefulness of the protocol.

Labeling oligonucleotides with digoxigenin (standard protocol is useful)

1. Mix
   - 4 μl of tailing buffer (vial 1)
   - 4 μl of $CoCl_2$ solution (vial 2)
   - 7 μl of distilled water
   - 100 pmol of oligonucleotide which has specific sequences for a certain mRNA
   - 1 μl of DIG-dUTP (vial 3)
   - 1 μl of dATP (vial 4)
   - 1 μl of terminal transferase with 50 unit/μl (vial 5)
2. Incubate for 15 mins at 37°C
3. Place the tube on ice and add 1 μl glycogen solution (vial 9) with 200 μl of 0.2 M EDTA, pH 8.0 (do not use phenol/chloroform in this step).
4. Precipitate the labeled oligonucleotide with 2.5 μl 4M LiCl and 75 μl prechilled ethanol. Mix well and leave for at least 30 mins at -70°C or 2 hrs at -20°C.
5. Centrifuge at 12,000 rpm, drain the supernatant and wash the pellet with 50 μl of 70% cold ethanol. Dry and dissolve in an appropriate volume of sterile water. Store at -20°C until use.

Hybridization

1. Take out the tightly sealed box, where the sections were stored, and warm it to 37°C for 60 mins.
2. Rehydration: soak the sections in PBS for 3 mins.
3. Fixation: fix the sections with 4% PFA in PBS for 15 mins at room temperature.
4. Washing: Wash the sections three times with PBS for 3 mins each time.
5. Deproteinization.

Note: This step increases the accessibility of the probe to the target mRNA in the tissue. The optimal treatment of the section with HCl and protease varies from one tissue to another and depends on the type of sample and condition of fixation. The optimal condition should be determined by each researcher. No requirement of this step is mentioned in some protocols when the length of oligonucleotide is less than 100 bases, however, in our experience, nuclei are often stained without deproteinization as false positive. React the sections with 0.2N HCl diluted with distilled water for 15 mins at room temperature. Wash sections in PBS three times. Treat the sections with 0.5 mg/ml of proteinase K diluted by PBS for 10 mins.

Note: Prewarm this diluted proteinase K solution for at least 30 mins before use. Otherwise, the reaction may not be enough when the whole solution is not warmed at 37°C.

6. Wash the sections in PBS three times.
7. Post fixation: Fix the sections in 4% PFA in PBS again for 10 mins at room temperature. Wash them in PBS three times.

Note: Sometimes after washing, treat the sections with 2 mg/ml of glycine in PBS for 15 mins twice in order to neutralize aldehyde completely. In our protocol, treatment with glycine is not performed.

8. Prehybridization: Soak the sections in the prehybridization buffer for at least 30 mins.

Note: It does not matter even if the duration of the above step is longer than 30 mins.

Note: Prehybridization may not be necessary in other protocols. However, during the preparation for hybridization, it is probably better to put the sections into a prehybridization buffer than to leave them in PBS.

9. Boiling the probe in the hybridization buffer: After adjustment of the probe concentration by the prehybridization buffer, put the solution in the 1.5 ml of eppen-tube and boil it for 5 mins. Put the tube into ice and water immediately and leave it for at least 10 mins.

Note: Immediate cooling es essential.

10. Hybridization: React the sections in the hybridization buffer at 37°C overnight.

Note: Reaction temperature varies depending on the Tm value (the length of the oligonucleotide and the degree of GC content).

Note: Seal the box tightly. Otherwise, the sections will dry.

Washing and detection of the probe (2nd day)

1. Drain the hybridization buffer on the glass and put it into washing solution 1(WS1). Wash the sections for 30 mins. Repeat the last step again.
2. Wash the sections twice in WS2 for 30 mins each.

Note: Determination of the washing condition is important for obtaining a good signal. If the washing condition is not tight (low stringency), the degree of background is higher. Conversely, if it is too tight (high stringency), the specific signal also becomes weaker with a low background level.

Note: From this step on, it is not necessary to be careful about contamination of RNase.

Visualization of the probe bound to specific mRNA on the tissue

Using immunohistochemistry, the probe bound to the specific mRNA in the tissue can be visualized using a light microscope. In our protocol, three-layer immunoperoxidase method with anti-DIG antibody and HRP-conjugated antibodies is performed to amplify the signal.

Note: The nonradioactive in situ hybridization method is generally considered to be less sensitive than the radioactive one. Our three-layer immunoperoxidase system is essential for the detection of hybridized probe. It remarkably increases the sensitivity of in situ hybridization using DIG-labeled oligonucleotide probe. Furthermore, in our original method, a four-layer immunoperoxidase system was used for the detection of the probe to enhance the sensitivity (Miyazaki et al. 1994).

1. Preincubation: After washing, put the preincubation buffer on the slide and leave it for at least 60 mins at room temperature.

Note: Do not dry the sections, otherwise non-specific signal develops.

2. Reaction with the anti-DIG antibody: Drain the preincubation buffer and react the sections with the monoclonal anti-DIG antibody diluted by the dilution buffer (final concentration; 1 μg/ml) for at least 60 mins at room temperature. Wash by PBS for 10 mins four times.

Note: HRP-conjugated sheep anti-DIG (Fab fragment) is also available from Boehringer, however, in immunohistochemistry this antibody is much less sensitive.

3. Methanol blocking: In order to block endogenous peroxidase, the sections are reacted with methanol containing 0.3% H_2O_2 for 15 mins at room temperature.

Note: In this step, sections must be shielded. After methanol blocking, wash the sections three times in PBS, each for 3 mins.

4. Reaction with second and third antibodies: Sections are reacted with one hundred times diluted HRP-conjugated rabbit anti-mouse immunoglobulin antibody for 30 mins at room temperature. After washing three times with PBS containing

0.075% BRIJ, each for 3 mins, sections are reacted with 50 times diluted HRP-conjugated swine anti-rabbit antibody for 30 mins at room temperature.

Note: Washing buffer containing BRIJ has a stronger effect, leading to a low background. Tween 20 instead of BRIJ can also be used.

5. Visualization: After washing four times with PBS containing 0.075% BRIJ, each for 3 mins, sections are reacted 0.5 mg/ml of 3,3'-diaminobenzidine tetrahydrochloride with H_2O_2. The duration of the reaction time in this step depends on the degree of color development (usually from 3 to 30 mins). After washing under running tap water, sections are stained with methyl green or hematoxyline-eosin, if nuclear staining is necessary. After treatment of the sections with serial concentrations of ethanol and with xylene, sections are mounted.

Note: If alkaline phosphatase (ALP) -conjugated antibody is used, DIG nucleic acid detection kit (Boehringer) is useful. The signal is dark green-gray in color. This system is very sensitive.

Required solutions for detection with ALP-conjugated antibody

- Buffer 1: maleic acid, 0.1 mol/l; NaCl, 0.15 mol/l; pH 7.5
- Blocking solution: Blocking reagent (vial 6) is dissolved with 10% (w/v) solution and stored at 4°C or -20°C.
- Buffer 2: Blocking solution is diluted to 1% with Buffer 1.
- Washing buffer: 0.3% Tween-20 in Buffer 1.
- Buffer 3: Tris-HCl, 100 mmol/l; NaCl, 100 mmol/l; $MgCl_2$, 50 mmol/l; pH 9.5.
- Buffer 4: Tris-HCl, 10 mmol/l; EDTA, 1 mmol/l; pH 8.0
- Color substrate solution (freshly prepared before use): Add NBT-solution (vial 4) and 35 μl X-phosphate-solution (vial 5) into 10 ml of Buffer 3. When blocking of endogenous alkali phosphatase is necessary, 10 drops of levamisole (Dako X3021) are added to the color substrate solution.

Immunological detection with ALP-conjugated anti-DIG antibody after hybridization followed by washing

1. Sections are reacted with Buffer 2 for 30 mins at room temperature.
2. After draining Buffer 2, sections are incubated with 5,000 times diluted alkali-phosphatase conjugated anti-DIG antibody by Buffer 2 for 60 mins.
3. Washing the sections by Buffer 1.
4. Treat the sections with Buffer 3 for 5 mins.
5. Sections are reacted with color substrate solution. Examine the signal intensity using a low power magnification, and stop the color development by washing with Buffer 4.
6. Wash the sections under running water for a short period of time then dry. Mount in Glycergel (Dako C0563).

Evaluation of specificity of the signal

1. Competitive study: A large amount of unlabeled oligonucleotide (50 - 200 times) is added to the standard hybridization mixture. Binding of DIG-labeled oligonucleotide to the target mRNA is inhibited and the signal becomes weak or disappears.
2. Sense probe study: Sense probe, which has a complementary sequence to anti-sense probe, is used instead of anti-sense probe. The signal is not detected.
3. RNase treatment: The section is treated with RNase before hybridization. The hybridization signal is markedly reduced, indicating that the signal is mRNA-related.

Note: The sequence of probe should be specific to the target mRNA. Any similarity between the sequence of the probe and that of any molecules should be checked by computer search of the GeneBank.

Note: When double strand DNA is used as a probe, it could be labeled with DIG DNA labeling kit (Boehringer) according the standard protocol. A template DNA is necessary (Should be prepared by each researcher). Preparation of the tissue, hybridization and immunohistochemistry after hybridization are similar to the methods described above. The protocol using double strand DNA probe is described in → Chapter 9.

Results

Figure 1 shows the expression of mRNA for interleukin (IL)-4 and IL-4 receptor in glomeruli of a representative patient with IgA nephropathy (Furusu et al. 1997). Note the clear staining of intraglomerular cells for IL-4 mRNA and IL-4 receptor mRNA in perinuclear or cytoplasmic pattern. Examination of mirror sections showed that some cells are double positive for IL-4 mRNA and IL-4 receptor mRNA, while others are positive for each mRNA. As shown in the figure, the expression

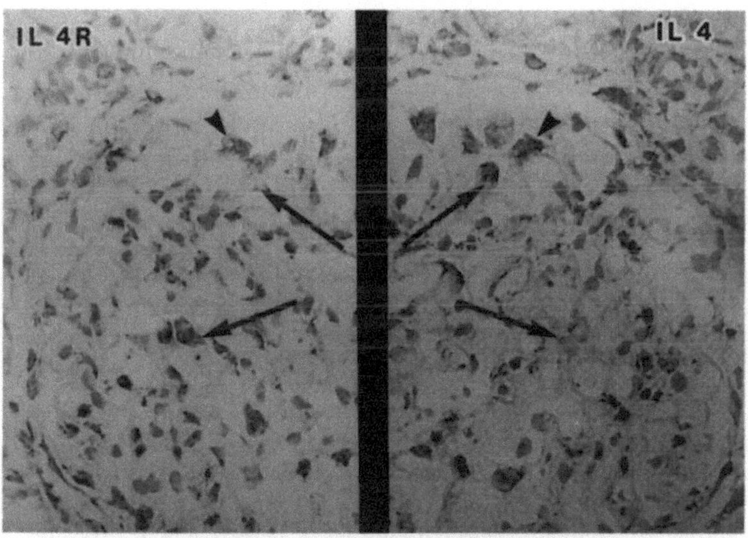

Fig. 1. In situ detection of IL-4 mRNA and IL-4 receptor mRNA in a renal tissue sample from a representative patient with IgA nephropathy. Mirror sections of glomeruli show that intraglomerular cells are clearly stained for IL-4 mRNA and IL-4 receptor mRNA in perinuclear or cytoplasmic pattern. As shown here, some cells are double positive for IL-4 mRNA and IL-4 receptor mRNA (arrowheads), while others are positive for each mRNA (arrows). The expression of the mRNAs is detected at a cellular level. Magnification x400.

of specific mRNA is detected at a cellular level. Our protocol has proved to be useful for clarifying the expression of mRNA for a variety of molecules in renal tissues, including, mRNA for complement and its inhibitor (Abe et al. 1996; Sasaki et al. 1997), extracellular matrix and metalloproteinase (Miyazaki et al. 1995). Our protocol can be modified from a 2-day method to 3-day one. In that case, sections are incubated with monoclonal anti-DIG antibody at 4°C overnight (the concentration of the antibody should be less than that in the 2-day method. On the third day, immunohistochemistry is performed as described above.

Acknowledgements

The authors thank Drs. A Furusu, K Abe and S Tsukasaki from the Second Department of Medicine of Nagasaki University and Drs. R Naka and D Suzuki from Department of Nephrology in Tokai University for their help. We are grateful to Professor H Sakai of Tokai University for his encouragement and support and Dr. PK Nakane, ex- Professor, Department of Anatomy, Nagasaki University for his advice. Dr. Nikolic-Paterson DJ and Professor RC Atkins at Monash Medical Center, Melbourne, Australia for helping us establish the original protocol of in situ hybridization. We are also grateful to Dr. Takehiko Koji for his indispensable suggestions.

References

Abe K, Miyazaki M, Koji T, Furusu A, Tsukasaki S, Ozono Y, Harada T, Nakane PK, Yagame M, Endoh M, Nomoto Y, Sakai H, Kohno S (1996) Expression of C3 mRNA and DAF mRNA in human diseased kidney. J Am Soc Nephrol 7: 1769.

Furusu A, Miyazaki M, Koji T, Abe K, Ozono Y, Harada T, Nakane PK, Hara K, Kohno S (1997) Involvement of interleukin-4 in human glomerulonephritis. -An in situ hybridization study for IL-4 mRNA and IL-4 receptor mRNA- J Am Soc Nephrol 8:730-741.

Koji T, Nakane PK (1990) Localization in situ of specific mRNA using thymine-thymine dimerized DNA probes. Sensitive and reliable non-radioactive in situ hybridization. Acta Pathol Jpn 40: 793-807.

Koji T and Nakane PK (1996) Recent advances in molecular histochemical techniques: In situ hybridization and Southwestern histochemistry. J Electron Microsc 45:119-127

Miyazaki M, Nikolic-Paterson DJ, Endoh M, Nomoto Y, Sakai H, Atkins RC, Koji T (1994) A sensitive method of non-radioactive in situ hybridization for mRNA localization within human renal biopsy specimens: use of digoxigenin labeled oligonucleotide. Internal Med 33: 87-91.

Miyazaki M, Koji T, Furusu A, Abe K, Ozono Y, Harada T, Naka R, Suzuki D, Yagame M, Endoh M, Nomoto Y, Hara K, Nakane PK, Sakai H (1995) In situ hybridization studies of stromelysin and tissue inhibitor of metalloproteinase 1 in IgA nephropathy. Nephrology 1: 119-127.

Miyazaki M, Abe K, Koji T, Furusu A, Ozono Y, Harada T, Nakane PK, Yagame M, Endoh M, Nomoto Y, Sakai H (1996) Intraglomerular C3 synthesis in human kidney detected by in situ hybridization. J Am Soc Nephrol 7: 2428-2433.

Sasaki O, Zhou W, Miyazaki M, Abe K, Koji T, Verroust P, Tsukasaki S, Ozono Y, Harada T, Nakane PK, Kohno S, Sacks SH (1997) Intraglomerular C3 synthesis in rats with passive Heymann nephritis. Am J Pathol 151: 1249-1256.

Schulte EK, Lyon HO, Hoyer PE (1992) Simultaneous quantification of DNA and RNA in tissue sections. A comparative analysis of the methyl green-pyronin technique with the gallocyanin chromalum and Feulgen procedures using image cytometry. Histochem J 24: 305-310.

Suzuki D, Miyazaki M, Naka R, Koji T, Yagame M, Jinde K, Endoh M, Nomoto Y, Sakai H (1995) In situ hybridization of interleukin 6 in diabetic nephropathy. Diabetes 44: 11233-1238.

Yoshii A, Koji T, Ohsawa N, Nakane PK (1995) In situ localization of ribosomal RNAs is a reliable reference for hybridizable RNA in tissue sections. J Histochem Cytochem 43:321-327.

In Situ Hybridization for RNA: Nonradioactive Probe: RNA Probe

HARUO OHTANI

Introduction

Non-radioactive in situ hybridization method is now widely used to detect nucleic acid in tissue sections. However, the results for the detection of tissue mRNA are not always consistent. We have to consider several factors for this purpose. We describe our current methods for the most consistent detection.

Procedure

Preparation of template DNA

1. The vector should contain promoter regions for RNA polymerases (for example, SP6 + T7 or T3 + T7).
2. The plasmid containing cDNA should be linealized with an appropriate restriction enzyme, and should be purified by ethanol precipitation.
3. The procedures should be RNase-free after phenol extraction.

In vitro transcription

Transcribe digoxigenin-labeled cRNAs of both anti-sense and sense orientations *in vitro* as follows.

Haruo Ohtani, Tohoku University School of Medicine, Department of Pathology, 2-1 Seiryo-machi, Aoba-ward, Sendai, 980-8575, Japan

template DNA (1μg/μl)	1μl
10x NTP mixture	2μl
10x transcription buffer	2μl
nuclease-free water	12μl
RNase inhibitor	1μl (20U)
Vortex	
Add RNA polymerase	2μl (40U)
Tip the tube, and incubate at 37 °C for 2hrs.	

Confirm the reaction by gel electrophoresis.

Purification of cRNAs

1. Add 1μl of 0.5M EDTA, 2μl of 4M LiCl, 1μl of 10mg/ml E. coli tRNA, 75μl of prechilled 100% ethanol. Mix well.
2. Stay at -70 °C for 2hrs or at -20 °C overnight. Microfuge at 14,000rpm for 15 mins. Discard the supernatant. Rinse the precipitate with 70-80% ethanol. Dry up the precipitate.

Sizing of cRNAs

1. Reduce the length of cRNAs to 100-150 bases to so that they can permeate tissue sections.
2. Add the following in the tube:
   - 95μl of alkaline solution (63mM Na_2CO_3+42mM $NaHCO_3$)
   - 5μl of 0.1M DTT.
3. Stay the cRNA solutions at 60°C. The reaction time (T) is calculated as follows: T (min) = $(L_0 - L_f)/KxL_0xL_f$. L_0, length (kb) of cRNA before treatment. Lf, length (kb) of cRNA after treatment. K=0.11

4. Add 10μl of 3M sodium acetate, 5μl of 20mg/ml and E. coli tRNA and 350μl of chilled ethanol. Stay at -70 °C for 2hrs or at -20 °C overnight. Microfuge at 14,000rpm for 15 mins. Rinse the precipitate with 70-80% ethanol. Dry up. Dissolve the precipitate with 50μl of nuclease free water (or TE buffer) containing 20U RNase inhibitor. Store at -20 or -70°C.

Quantification of the cRNAs by immunohistochemical staining

1. Dilute the solution at 1:100, 1:200, 1:400, 1:800, and 1:1600. The dilution buffer is made by adding 5μ l of 10mg/ml E. coli tRNA into 1000μl of TE buffer.
2. Mark them to the filter membrane (1μl) and crosslink them by ultraviolet.
3. Stain the filter with anti-digoxigenin antibody, and develop the color reagents by alkaline phosphatase substrate.
4. Compare each reaction with that of control labeled RNA.

Fixation

Both paraffin-embedded sections and frozen sections can be used.

The most frequently used fixative is 4% paraformaldehyde solution (PFA)/phosphate buffer. The fixation time is 2hrs to overnight at 4°C.

Our current fixatives are 4%PFA + 0.5% glutaraldehyde/PBS (0.075M, pH 7.2) containing 1mM $MgCl_2$ and 1mM $CaCl_2$. This fixation is quite useful because of higher sensitivity and more consistent results[1,2]. Immerse sliced tissues in this fixative overnight at 4 °C. The thickness of tissue should be 2mm. If thick (›2mm) sections were fixed, uneven fixation results will be gained. If possible, perfusion fixation is more preferable. This fixative is useful for the fixation of bone tissue followed by EDTA decalcification[3].

After the completion of fixation, the specimens are either dehydrated by ethanol to be embedded in paraffin or frozen. For frozen sections, the specimens are washed in 20-30% sucrose/

PBS for about 2 hrs, embedded in an embedding substance (for example O.C.T. compound), and frozen in dry ice/acetone.

Cut sections, mount them on silane-coated glass slides.

If this fixative is used, there are no significant differences of sensitivity between paraffin-embedded sections and that of frozen sections.

Hybridization (paraffin-embedded sections)

Prehybridization (RNase-free condition)

1. Deparaffinize sections in xylene-ethanol series.
2. Heat the sections in 10mM citrate buffer (pH 6.0) by microwave (15 mins after boiling).
3. Wash the sections in PBS.
4. Treat the sections with proteinase K (20µg/ml in PBS) for 30 mins at 37 °C.
5. Wash the sections in PBS, 3x.
6. Fix the sections in 4% PFA in PBS for 5 mins.
7. Immerse the sections in glycine (10mg/ml) for 15 mins, 2x.
8. Immerse the sections in 2x SSC/50% deionized formamide, 30mins

The procedure step 6-8 can be replaced by dehydration by ethanol and dry-up. The heating procedure (step 2) is used to enhance the sensitivity (mildly effective).

For frozen sections, first cut frozen sections, mount them onto coated-glass slides, and air-dry them for 1hr. When fresh frozen sections are used, fix tissue sections with 4% PFA for 30 mins. Immerse sections in PBS to remove OCT compound. The hybridization procedures are the same as those of paraffin-embedded sections except that the condition of proteinase K treatment is mild; i.e., 2µg/ml for 10 mins at 37 °C.

Hybridization (RNase-free condition)

Add the following (for 1000μl):

20mg/ml E. coli tRNA	50μl
1.0M Tris HCl buffer, pH8.0	20μl
0.5M EDTA, pH8.0	5μl
20x Denhart's solution	50μl
5M NaCl	60μl
deionized formamide	500μl
DEPC-treated water	115μl
50% dextran sulfate	200μl
digoxigenin-labeled cRNA	an appropriate volume

Note: The final concentration of the probes is set at 1μg/ml at first. If the reaction is too strong, dilute the probe concentration. Vortex well. Heat the solution at 95°C for 5 mins and cool it on ice.

Apply the hybridization solution on the glass slides. The quantity of the hybridization solutions is 20-80μl per one section depending on the size of the specimens. Incubate the sections at 45°C for 16hrs.

The cover glass is not required. On those occasions, the moist chamber used for the hybridization should contain 2x SCC/50% formamide (FA) to avoid drying or excessive hydration of hybridization solution applied on the specimens.

The glassware and solutions should be RNase free till this process. It may be difficult to make the moist chamber RNase-free. In this case, the chambers should be carefully washed each time after use.

Posthybridization washing

Use conventional glassware and solutions (no need for RNase-free procedures after this).

1. Wash in 2x SSC/50% FA for 1h at 50°C.

Note: The temperature can be set at 45-50°C in the washing procedures. This should be set higher when non-specific reactivity is observed.

2. Wash sections in 2x SSC, 2x.
3. Treat the sections in RNase A (20μg/ml) in 2x SSC for 30 mins at 37°C.
4. Wash sections in 2x SSC/50%FA for 1 hr at 50°C.
5. Wash sections in 1x SSC/50%FA for 1hr at 50°C.

Immunohistochemical detection of signals

1. Wash sections in PBS, 3x.
2. Immerse the sections in the blocking solution to block the non-specific binding of antibody to the sections.
3. Apply anti-digoxigenin antibody conjugated with alkaline phosphatase (1:500, 1hr-overnight) on the sections and incubate them for 1hr to overnight.

Note: Horseradish peroxidase can also be used.

4. Wash the sections with Tris HCl (10mM, pH 7.6)-buffered saline (or PBS) containing 0.3% Tween 20, 3x.
5. Wash the sections with 100mM Tris HCl buffer (pH 9.5)/ 100mM NaCl/50mM MgCl2 (phosphate should be removed).
6. Apply coloring solution (NBT/BCIP) on the sections. Put the sections in a moist chamber in the dark. The reaction products appear in 30 mins to overnight.
7. Wash the sections in running water.
8. Fix the sections in 10% formalin for 10 mins.
9. Wash the sections in PBS for 30 mins to overnight. This may be effective to avoid the precipitation of crystals on the reaction products.
10. Counterstain the sections with 2% methyl green for 10 mins.
11. Wash the sections with running water.
12. Air-dry the sections.
13. Drop xylene, and mount the sections with an embedding substance.

Note: The original mounting method is aqueous mounting. However, permanent specimens can be made by this method.

Negative control The negative control with cRNA of a sense orientation is important for the conformation of the specificity.

If the reaction products are the same between cRNAs of both sense and antisense orientations, the following negative control ("competitive inhibition assay") can be used.

Transcribe non-labeled cRNA of antisense orientation as shown below.

Add (with the total volume of 100μl)

Note: 10-fold larger amount of template DNA or RNA polymerase is not required for this reaction.

template DNA (1μg/μl)	2.5μl
10x transcription buffer	10μl
0.1M DTT	10μl
10mM rATP	5μl
10mM rCTP	5μl
10mM rGTP	5μl
10mM rUTP	5μl
nuclease-free water	53μl
RNase inhibitor	2.5μl
RNA polymerase	2μl

Tip the tube, incubate for 2hrs at 37°C.

Ethanol precipitation and sizing should be done in the same way.

Do not add tRNA in the whole process. Measure RNA concentration at OD_{260}.

The negative control is done by adding 100-fold larger amount of non-labeled cRNA in the reaction solution containing labeled cRNA with an antisense orientation.

If the reactivity is completely abolished, then further control is to be done to confirm whether this abolishment of the reaction is specific or not by adding the same amount of non-corresponding, non-labeled RNA. When this control does not affect the reactivity, then the results with the digoxigenin-labeled cRNA with an anti-sense orientation are specific even though the labeled cRNA with a sense orientation showed reactivity.

General advice for this method

- Prolonged incubation time for alkaline phosphatase reaction tends to yield non-specific reaction products. These should not be taken as specific. Always be careful with the results of negative control. In our experience the possibility of non-specific reactivity is higher by overnight incubation of alkaline phosphatase reaction.
- Proper fixation is preferable to avoid non-specific reactivity, since the reaction time of alkaline phosphatase will be shortened considerably with a clear reactivity by adopting proper fixation of, for example, 4% paraformaldehyde + 0.5% glutaraldehyde.
- Therefore, caution is required when one uses formalin-fixed, paraffin-embedded conventional blocks. The author recommends using properly fixed blocks.

References

Motohashi H, Igarashi K, Onodera K, Takahashi S, Ohtani H, Nakafuku M, Nishizawa M, Engel JD, Yamamoto M. Mesodermal- vs. Neuronal-specific expression of MafK is elicited by different promoters. Genes Cells 1996;1:223-238.

Ohtani H, Motohashi H, Sato H, Seiki M, Nagura H. Dual overexpression pattern of membrane-type metalloproteinase-1 in cancer and stromal cells in human gastrointestinal carcinoma revealed by *in situ* hybridization and immunoelectron microscopy. Int J Cancer 1996;68:565-570.

Ohtani H, Nagai T, Nagura H. Similarities of *in situ* mRNA expression between gelatinase A (MMP-2) and type I procollagen in human gastrointestinal carcinoma: comparison with granulation tissue reaction. Jpn J Cancer Res 1995;86:833-839.

Sasano Y, Furusawa M, Ohtani H, Mizoguchi T, Takahashi, T, Kagayama M. Chondrocytes synthesize type I collagen and accumulate the protein in the matrix during development of rat tibial articular cartilage. Anat Embryol 1996;194:247-252.

Electron Microscopic in Situ Hybridization and its Combination With Immunohistochemistry

AKIRA MATSUNO, TADASHI NAGASHIMA,
R. YOSHIYUKI OSAMURA AND KEIICHI WATANABE

Introduction

In situ hybridization at the electron microscopic level is a useful method for examining the intracellular distribution and role of mRNA in protein synthesis. Ultrastructural in situ hybridization for mRNA has been carried out by three different approaches: preembedding method (Guitteny et al. 1989; Wolber et al. 1989; Trembleau et al. 1990; Pomeroy et al. 1991; Matsuno et al. 1994a,b, 1995), ultrathin frozen sections (Morel et al. 1989a,b; Le Guellec et al. 1992) and postembedding method (Jirikowski et al. 1990; Le Guellec et al. 1990, 1991, 1992, Matsuno et al. 1994a, 1995). In recent years, we have developed non-radioisotopic electron microscopic in situ hybridization method using biotinylated synthesized oligonucleotide probes for the ultrastructural visualization of growth hormone (GH) and prolactin (PRL) mRNAs in rat pituitary cells, and have applied this method to pathophysiological studies of pituitary cells (Matsuno et al. 1994a,b, 1995). In addition, we have developed combined method of electron microscopic in situ hybridization and immunohistochemistry for simultaneous detection of pituitary hormone and its message in the same cell (Matsuno et al. 1996), which can

Correspondence to Akira Matsuno, Teikyo University Ichihara Hospital, Department of Neurosurgery, 3426-3 Anegasaki, Ichihara City, Chiba, 299-0111, Japan (*phone* +81-436-62-1211; *fax* +81-436-62-1357; *e-mail* akirakun@med.teikyo-u.ac.jp); Tadashi Nagashima, Teikyo University Ichihara Hospital, Department of Neurosurgery, 3426-3 Anegasaki, Ichihara City, Chiba, 299-0111, Japan; R. Yoshiyuki Osamura, Tokai University School of Medicine, Department of Pathology, Boseidai, Isehara City, Kanagawa, 259-1100, Japan; Keiichi Watanabe, Tokai University School of Medicine, Department of Pathology, Boseidai, Isehara City, Kanagawa, 259-1100, Japan

serve the visualization of intracellular spatial relationship between pituitary hormone and its message at the ultrastructural level. In this chapter, both preembedding and postembedding methods of non-radioisotopic electron microscopic in situ hybridization are introduced together with combined immunohistochemistry and preembedding non-radioisotopic in situ hybridization technique.

Outline

The protocol of each of the following techniques is outlined in Fig. 1-3.

Preembedding non-radioisotopic electron microscopic in situ hybridization (Fig. 1)

Postembedding non-radioisotopic electron microscopic in situ hybridization (Fig. 2)

Combined immunohistochemistry and preembedding non-radioisotopic electron microscopic in situ hybridization (Fig. 3)

Frozen section fixed with 4% PFA

↓

Hybridization with biotinylated probe at 37℃ overnight

↓

Signal detection by ABC-HRP-DAB

↓

Osmification

↓

Embedment in Epon resin

↓

Electron microscopic observation

Fig. 1. Outline of preembedding electron microscopic in situ hybridization. PFA: paraformaldehyde, ABC-HRP-DAB: streptavidin-biotin complex, horseradish peroxidase and diaminobenzidine

Embedment of tissues fixed with 4% PFA in LR White resin

↓

Ultrathin section mounted on Nickel grid

↓

Hybridization with biotinylated probe at 37℃ overnight

↓

Signal detection by 20nm streptavidin gold

↓

Electron microscopic observation

Fig. 2. Outline of postembedding electron microscopic in situ hybridization.

Frozen section fixed with 4% PFA

↓

Hybridization with biotinylated probe at 37℃ overnight

↓

Signal detection by ABC-HRP-DAB

↓

Osmification

↓

Embedment in Epon resin

↓

Ultrathin sections mounted on Nickel grids

↓

Etching with 10% H2O2 or 4% sodium periodate

↓

Application of primary antibody

↓

Visualization with Protein-A colloidal gold

↓

Electron microscopic observation

Fig. 3. Outline of combined immunohistochemistry and preembedding electron microscopic in situ hybridization.

Materials

Equipment

- Electron microscopy (e.g., JEOL 1200EX)
- incubator
- vacuum oven

Reagents

Ribonuclease-free solutions treated with 0.02% diethylpyrocarbonate (DEPC), autoclaved glass ware and gloves should be used for in situ hybridization study.

- synthesized oligonucleotide probes
- ENZO's terminal labeling kit (ENZO Diagnostics Inc., Farmingdale, NY, USA)
- 4% paraformaldehyde
- 0.01 M phosphate buffered saline pH 7.4 (PBS)
- graded concentrations of sucrose dissolved in PBS (10%, 15%, 20%)
- Optimal Cutting Temperature (OCT) compound (Tissue-Tek; Miles Laboratories Inc., Elkhart, Ind., USA)
- 3-aminopropylmethoxysilane-coated slides
- proteinase K
- acetic anhydride
- triethanolamine
- 2x sodium chloride sodium citrate (SSC)
- prehybridization solution: 10% dextran sulfate, 3xSSC, 1x Denhardt's solution (0.02% Ficoll/ 0.02% bovine serum albumin (BSA)/ 0.02% polyvinylpyrrolidone), 100 μg/ml salmon sperm DNA, 125 μg/ml yeast tRNA, 10 μg/ml polyadenylic-cytidylic acid, 1 mg/ml sodium pyrophosphate pH 7.4 and 50% formamide
- Vectastain's ABC kit (Vector Laboratories Inc., Burlingame, Ca., USA)
- diaminobenzidine (DAB) solution: 20 mg of freshly prepared 3,3'-diaminobenzidine tetrahydrochloride dissolved in 0.05 M Tris-HCl pH 7.6
- H_2O_2
- osmium tetroxide
- Epon resin (e.g., Quetol-812)
- LR White resin (Polyscience ; Warrington, PA)
- beam capsule
- 00 gelatin capsules (Lilly Pharmaceuticals ; Indianapolis, IN)

- sodium periodate
- 0.1 M phosphate buffer (PB)
- 20 nm protein A colloidal gold (EY Laboratories Inc., San Mateo, Ca., USA)
- 20 nm streptavidin colloidal gold (Bio Cell Research Laboratories, Cardiff, UK)

Procedure

Preembedding method

1. Tissues are fixed at 4°C overnight in 4% paraformaldehyde dissolved in PBS.
2. Tissues are immersed in graded concentrations of sucrose dissolved in PBS at 4°C (10% for 1 hr, 15% for 2 hrs, 20% for 4 hrs), and are embedded in OCT compound.
3. Six μm thick tissue sections are mounted on 3-aminopropylmethoxysilane-coated slides.
4. After air drying for 1 hr, tissue sections are washed with PBS for 15 mins.
5. Tissue sections are treated with 0.1 μg/ml proteinase K in PBS at 37°C for 30 mins, followed by treatment for 10 mins with 0.25% acetic anhydride in 0.1 M triethanolamine.
6. The slides are washed in 2x SSC at room temperature for 3 mins and then prehybridized with prehybridization solution at 37°C for 30 mins.
7. The biotinylated probe with the concentration of 0.1 ng/μl is diluted with this prehybridization solution and hybridization is carried out at 37°C overnight.
8. The slides are washed at room temperature with 2xSSC, 1xSSC and then 0.5xSSC for 15 mins each.
9. After being washed in PBS for 5 mins, the slides are dipped in DAB solution for 1 hr.
10. The hybridization signals are detected with streptavidin-biotin-horseradish peroxidase, using Vectastain's ABC kit,

and thereafter developed with 0.017% H_2O_2 in DAB solution for 7 mins.

11. 2% osmium tetroxide is applied to the sections. After dehydration with graded ethanol (50%, 70%, 80%, 90%, 95%, 100%), tissue sections are embedded in Epon resin by inverted beam capsule method.

12. Polymerization at 60°C for 2 days. Ultrathin sections are inspected under an electron microscopy.

The negative control experiments: hybridization with probes of sense or scramble sequence and without probes.

Postembedding method

1. Tissues are fixed at 4°C overnight in 4% paraformaldehyde dissolved in PBS, and embedded in LR White resin. Tissues are carefully placed at the bottom of 00 gelatin capsules, which are filled with LR White resin and sealed.

2. After polymerization at 50°C for 24 hrs in a vacuum oven, ultrathin sections are retrieved on Nickel grids.

3. After prehybridization at 37 °C for 30 mins, hybridization is carried out on the grids at 37°C overnight. The concentration of the biotinylated oligonucleotide probe is 1 ng/μl.

4. After hybridization, the grids are dipped in 2xSSC, 1xSSC and then 0.5xSSC for 5 mins each.

5. The hybridization signals are developed for 30 mins with 20 nm streptavidin gold diluted 1:50 in 1% BSA-PBS. After being dipped in PBS and distilled water and dried at room temperature, the grids are inspected under an electron microscopy.

The negative control experiments: hybridization with probes of sense or scramble sequence and without probes.

Combined immunohistochemistry and preembedding in situ hybridization at an electron microscopic level

Steps 1 to 11 are the same as described in the section Preembedding Method.

12. Polymerization at 60°C for 2 days.
13. The ultrathin sections are attached to Nickel grids.
14. In order to retrieve the immunoreactivity of the targeted protein, the ultrathin sections embedded in Epon resin are etched with 10% H_2O_2 for 30 mins or with 4% sodium periodate for 10 mins, followed by washing with distilled water.
15. Immunohistochemical staining is carried out at room temperature for 1 hr using the appropriate antibody.
16. The grids are washed with 0.1 M PB, and the immunoreaction is visualized at room temperature for 1 hr with 20 nm protein A colloidal gold diluted 1:40 in 0.1 M PB.
17. After being washed with 0.1 M PB and distilled water, the grids are dried at room temperature, and observed under an electron microscopy.

The immunohistochemical negative control experiments: substitution of normal serum for the antibody.

Combined immunohistochemistry and preembedding in situ hybridization at an electron microscopic level using LR White resin

Steps 1 to 10 are the same as described in the section under Preembedding Method.

11. 2% osmium tetroxide is applied to the sections. After dehydration with graded ethanol (50%, 70%, 80%, 90%, 95%, 100%), tissue sections are embedded in LR White resin using inverted beam capsule method.
12. Polymerization at 50°C for 2 days in a vacuum oven.
13. The ultrathin sections are attached to Nickel grids.

14. Immunohistochemical staining is carried out at room temperature for 1 hr using the appropriate antibody.
15. The grids are washed with 0.1 M PB, and the immunoreaction is visualized at room temperature for 1 hr with 20 nm protein A colloidal gold diluted 1:40 in 0.1 M PB.
16. After being washed with 0.1 M PB and distilled water, the grids are dried at room temperature, and observed under an electron microscopy.

The immunohistochemical negative control experiments: substitution of normal serum for the antibody.

Results

In this section, the studies on the intracellular localization of GH and PRL mRNA in the rat pituitary cells, to which the above de-

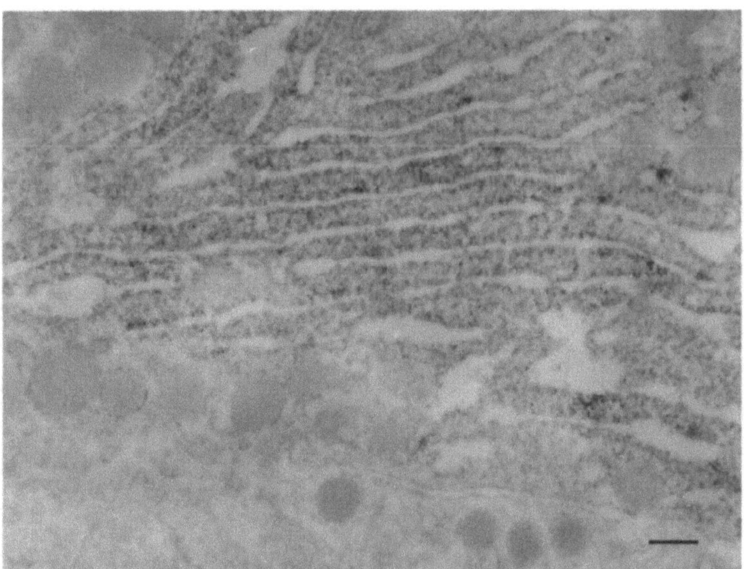

Fig. 4. Photograph of preembedding electron microscopic in situ hybridization for rat GH mRNA. Rat GH mRNA is localized diffusely on the polysomes of the entire rough endoplasmic reticulum. (bar=200nm) (Cited from Matsuno et al. 1994a)

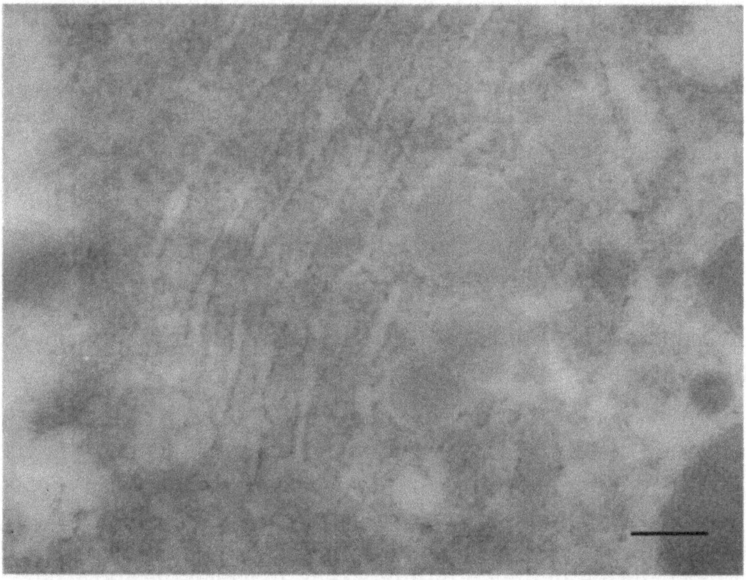

Fig. 5. Control study with a sense probe for rat GH mRNA yields no hybridization signals on the rough endoplasmic reticulum. (bar=200nm) (Modified from Matsuno et al. 1994a)

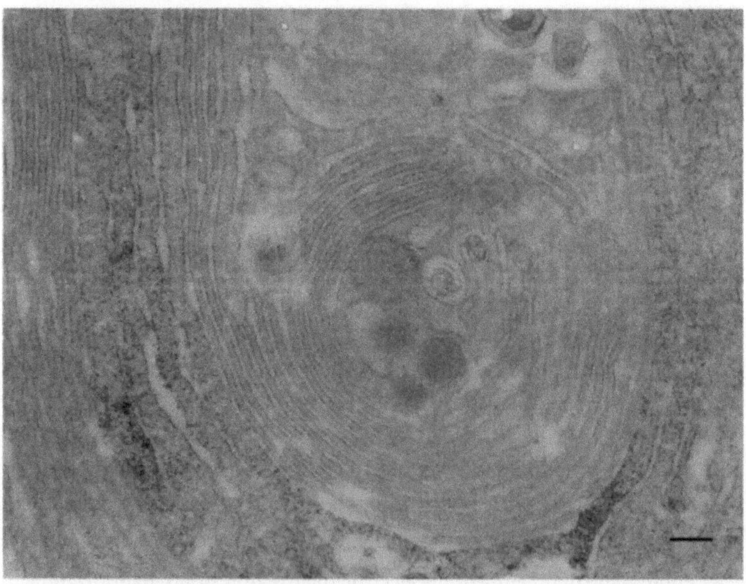

Fig. 6. Photograph of preembedding electron microscopic in situ hybridization for rat PRL mRNA in pituitary cells from female rats treated with estrogen. Rat PRL mRNA is scattered and localized focally on the whirling rough endoplasmic reticulum. (bar=200nm)

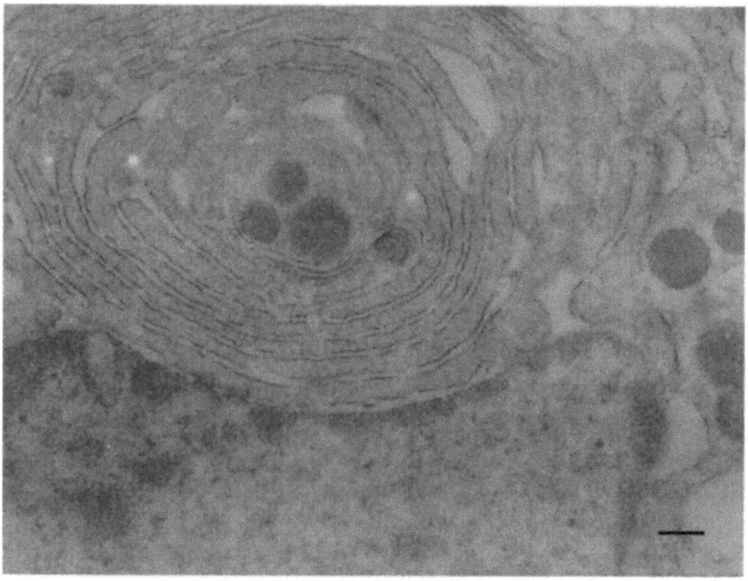

Fig. 7. Control study with a sense probe for rat PRL mRNA yields no hybridization signals on the rough endoplasmic reticulum. (bar=200nm)

scribed electron microscopic in situ hybridization method is applied, are illustrated.

Preembedding method

Electron microscopic in situ hybridization with an antisense probe for rat GH mRNA reveals its diffuse localization on the polysomes of the entire rough endoplasmic reticulum (Fig. 4). Electron microscopic in situ hybridization with a sense probe for rat GH mRNA shows negative signals (Fig. 5). In pituitary cells from female rats treated with estrogen, rat PRL mRNA is scattered and localized focally on the polysomes of the whirling rough endoplasmic reticulum (Fig. 6). Electron microscopic in situ hybridization with a sense probe for rat PRL mRNA shows negative signals (Fig. 7).

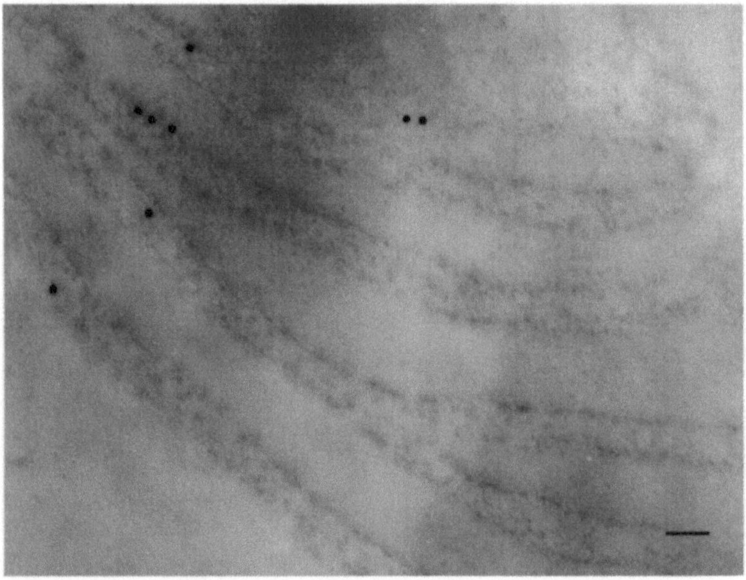

Fig. 8. Hybridization signals for rat GH mRNA are localized on the polysomes of the rough endoplasmic reticulum using 20 nm streptavidin gold. (bar=200nm) (Cited from Matsuno et al. 1994a)

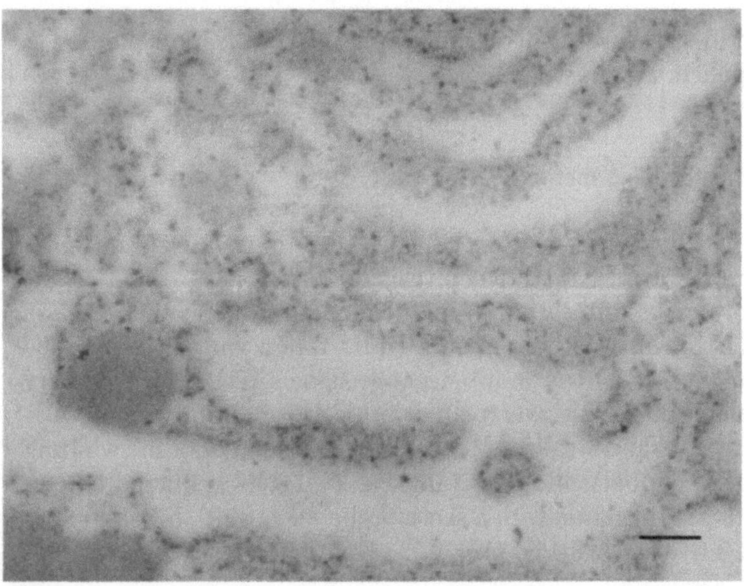

Fig. 9. Electron microscopic in situ hybridization with a sense probe for rat GH mRNA shows no hybridization signals. (bar=200nm) (Cited from Matsuno et al. 1994a)

Postembedding method

Hybridization signals for rat GH mRNA are localized on the polysomes of the rough endoplasmic reticulum using 20 nm streptavidin gold (Fig. 8). Electron microscopic in situ hybridization with a sense probe for rat GH mRNA shows no hybridization signals (Fig. 9).

Combined immunohistochemistry and preembedding in situ hybridization at an electron microscopic level

Electron microscopic in situ hybridization with an antisense probe for rat GH mRNA reveals its diffuse localization on the polysomes of the entire rough endoplasmic reticulum. Subsequent immunohistochemical staining using anti-rat GH anti-

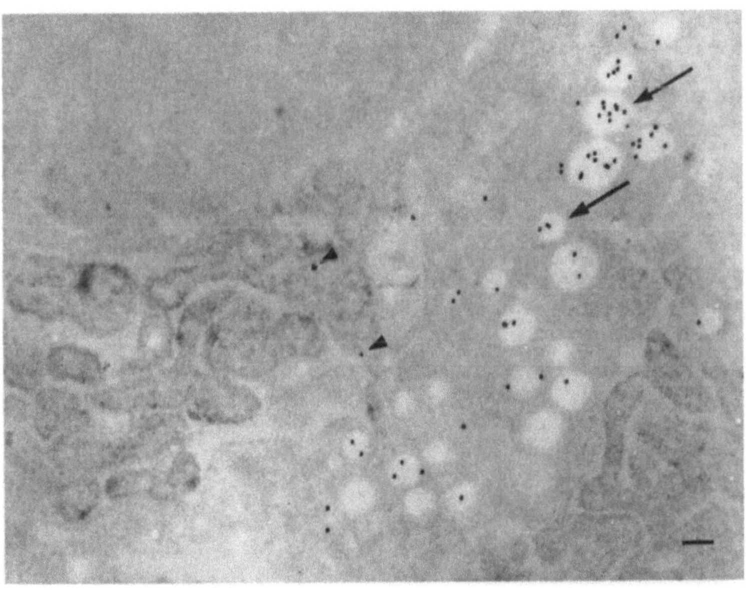

Fig. 10. Combined immunohistochemistry and preembedding electron microscopic in situ hybridization at the ultrastructural level. Positive signals for rat GH mRNA are localized as osmium black on the polysomes of the rough endoplasmic reticulum using biotinylated antisense oligonucleotide probe. Rat GH protein is identified mainly on the secretory granules with 20 nm protein A colloidal gold particles (arrows), and is also identified in the cisternae of the rough endoplasmic reticulum (arrow heads). (bar=200nm)

body and 20 nm protein A colloidal gold identifies rat GH protein on the secretory granules (Fig. 10). Colloidal gold signals for rat GH protein are distributed mainly on the secretory granules, and are also identified in the cisternae of the rough endoplasmic reticulum (Fig. 10). The immunoreactivity retrieved after etching process with 10% H_2O_2 is similar and comparable to that with 4% sodium periodate. Immunohistochemical control experiment combined with preembedding electron microscopic in situ hybridization method reveals no reactions of protein A colloidal gold particles on the secretory granules (Fig. 11).

Combined immunohistochemistry and preembedding in situ hybridization at an electron microscopic level using LR White resin

Using LR White resin for tissue embedment, rat GH mRNA is also localized on the polysomes of the entire rough endoplasmic

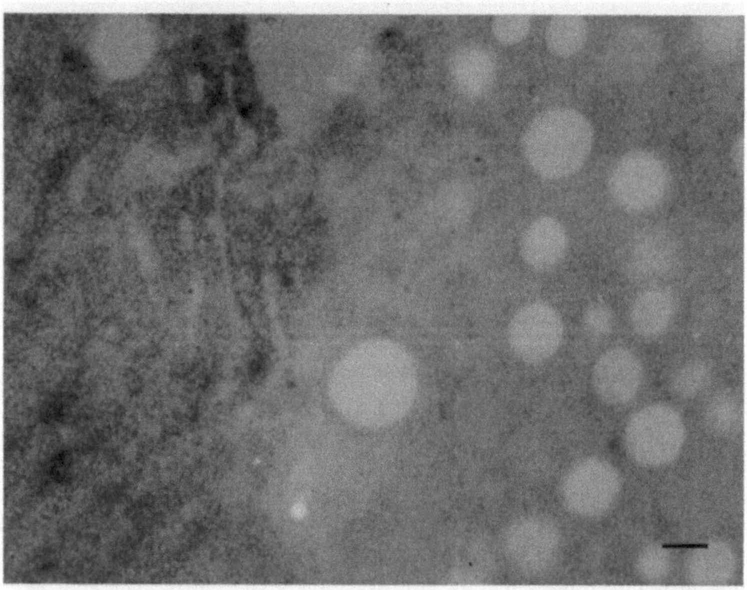

Fig. 11. Immunohistochemical control experiments at the ultrastructural level. No positive reactions of protein A colloidal gold particles are observed on the secretory granules, whereas positive signals for rat GH mRNA are identified as osmium black on the polysomes of the rough endoplasmic reticulum. (bar=200nm)

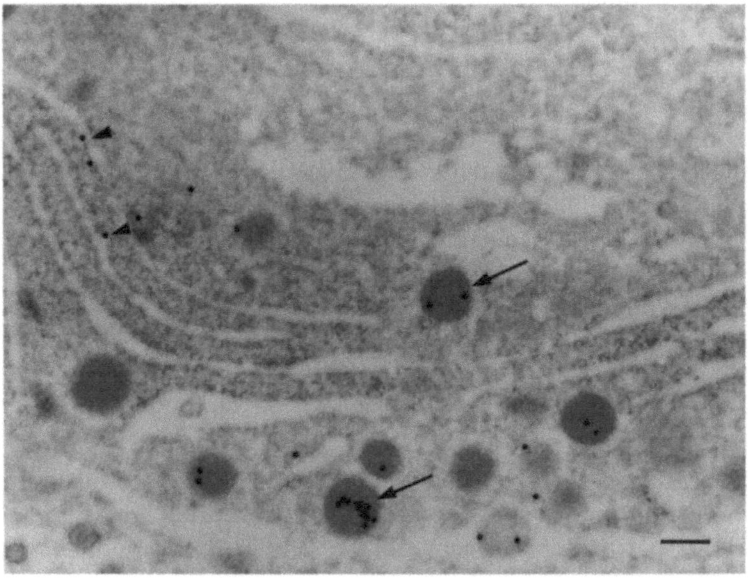

Fig. 12. Using LR White resin for tissue embedment, rat GH mRNA is also localized on the polysomes of the entire rough endoplasmic reticulum. Subsequent immunohistochemical staining identifies rat GH protein both on the secretory granules (arrows) and in the cisternae of the rough endoplasmic reticulum (arrow heads). (bar=200nm)

reticulum (Fig. 12). Subsequent immunohistochemical staining identifies rat GH protein both on the secretory granules and in the cisternae of the rough endoplasmic reticulum (Fig. 12).

Troubleshooting

- Prolonged fixation of the tissues will decrease the messages.
- Excessive treatment with proteinase K will destroy the morphology.
- Excessive etching process will shade off the reaction products (osmium black) of mRNA.

Comments

Light microscopic in situ hybridization, which has become a widely used method for examining the tissue distribution and expression of mRNA, lacks the ultrastructural spatial relationship between mRNA and the encoded protein. This information, which can be obtained by electron microscopic in situ hybridization, will be needed to elucidate the intracellular distribution and role of mRNA for protein synthesis. The major concerns of electron microscopic in situ hybridization are to maintain tissue morphology and retain the messages. As Le Guellec et al. stated (Le Guellec et al. 1992), the ultrastructural preservation in electron microscopic in situ hybridization using ultrathin frozen sections is poor, and specimens embedded in Lowicryl K4M exhibit poorer ultrastructural preservation than those embedded in Epon resin. In order to obtain satisfactory morphological preservation, we have routinely utilized 6 μm frozen sections fixed in 4% paraformaldehyde for the preembedding method and tissues embedded in LR White resin for the postembedding method.

As described in our previous reports (Matsuno et al. 1994a), compared with the preembedding method, the postembedding method presents several difficulties: 1) Difficulty of message preservation during polymerization of LR White resin at high temperature for an extended period of time, which leads to mRNA degradation. 2) Relatively high frequency of non-specific signals because of the non-specific affinity of gold particles. From these results, it is suggested that the preembedding method would be better and easier than the postembedding method because of its higher sensitivity and better preservation of mRNA.

Our other concern is to visualize mRNA and encoded protein simultaneously in the same cell. There seem to be two major problems to be resolved in this ultrastructural double staining method. One is to retain the messages and the other is to maintain the immunoreactivity of the encoded protein in the same cell. As for mRNA preservation the preembedding electron microscopic in situ hybridization method using frozen sections fixed in 4% paraformaldehyde has more advantages over the postembedding electron microscopic in situ hybridization method using tissues embedded in LR White resin. Frozen sections fixed in 4% paraformaldehyde have better morphological preservation than immediately frozen sections. Based on this assessment of maintaining tissue morphology and retaining the messages, we utilize the preembedding electron microscopic in situ hybridization method using 6 μm frozen sections fixed in 4% paraformaldehyde for the simultaneous detection of mRNA and encoded protein.

In general, osmification and embedment in Epon resin are reported to decrease the immunoreactivity of the targeted protein. However, immunoreactivity can be retrieved by the etching process using H_2O_2 or sodium periodate even after modification such as osmification and embedment in Epon resin, and tissues embedded in Epon resin can be served for the ultrastructural simultaneous detection of messages and encoded proteins. The only problem is the deosmification and degradation of the signals of mRNA, which are caused by etching process using H_2O_2 or sodium periodate. In order to resolve this problem, we have recently used LR White resin for tissue embedment (Matsuno et al. 1998). In LR White resin-embedded tissues, retrieval of immunoreactivity using H_2O_2 or sodium periodate is not required, and therefore the degradation of the signals of mRNA can be avoided.

To our knowledge, only five reports describing the ultrastructural simultaneous detection of mRNA and encoded protein have been published, in each of which postembedding electron microscopic in situ hybridization method using colloidal gold particle was utilized. Some investigators (Singer et al. 1989; Escaig-Haye et al. 1992; Egger et al. 1994; Gingras & Bendayan 1995) utilized Lowicryl K4M-embedded tissues for the detection of messages, while others (Morey et al. 1995) utilized LR White-embedded tissues. However, in their postembedding electron microscopic in situ hybridization studies, the relatively frequent non-specific reactions of colloidal gold particles used for the detection of mRNA were observed in the cisternae of the rough endoplasmic reticulum. Electron microscopic in situ hybridization method for Lowicryl K4M-embedded tissues is generally supposed to present some difficulty in morphological preservation. From the viewpoint of specific hybridization signal detection and morphological preservation, combined immunohistochemistry and preembedding electron microscopic in situ hybridization method is supposed to be superior to the previously reported postembedding electron microscopic in situ hybridization method. This ultrastructural double staining method for mRNA and encoded protein is considered to provide an important clue for elucidating the intracellular correlation of mRNA translation and secretion of translated protein.

The authors thank Susumu Takekoshi, Hirotoshi Utsunomiya, Yoshitaka Ohsugi and Shigeru Munakata for their dedicated assistance. This study was supported in part by the Grant-in-aid for Scientific Research from the Ministry of Education, Science and Culture, the Japanese Government.

References

Egger D, Troxier M, Bienz K (1994) Light and electron microscopic in situ hybridization: non-radioactive labeling and detection, double hybridization, and combined hybridization-immunocytochemistry. J Histochem Cytochem 42:815-822

Escaig-Haye F, Grigogiev V, Sharova I, Rudneva V, Buckrinskaya A, Fournier JG (1992) Ultrastructural localization of HIV-1 RNA and core proteins. Simultaneous visualization using double immunogold labelling after in situ hybridization and immunocytochemistry. J Submicrosc Cytol Pathol 24:437-443

Gingras D, Bendayan M (1995) Colloidal gold electron microscopic in situ hybridization: combination with immunocytochemistry for the study of insulin and amylase secretion. Cell Vision 2:218-225

Guitteny AF, Bloch B (1989) Ultrastructural detection of the vasopressin messenger RNA in the normal and Brattleboro rat. Histochemistry 92:277-281

Jirikowski GF, Sanna PP, Bloom FE (1990) mRNA coding for oxytocin is present in axons of the hypothalamo-neurohypophyseal tract. Proc Natl Acad Sci USA 87:7400-7404

Le Guellec D, Frappart L, Desprez PY (1991) Ultrastructural localization of mRNA encoding for the EGF receptor in human breast cell cancer line BT20 by in situ hybridization. J Histochem Cytochem 39:1-6

Le Guellec D, Frappart L, Willems R (1990) Ultrastructural localization of fibronectin mRNA in chick embryo by in situ hybridization using ^{35}S or biotin labeled cDNA probes. Biol Cell 70:159-165

Le Guellec D, Trembleau A, Pechoux C, Gossard F, Morel G (1992) Ultrastructural non-radioactive in situ hybridization of GH mRNA in rat pituitary gland: preembedding vs ultrathin frozen sections vs postembedding. J Histochem Cytochem 40:979-986

Matsuno A, Ohsugi Y, Utsunomiya H, Takekoshi S, Munakata S, Nagao K, Osamura RY, Tamura A, Nagashima T (1998) An improved ultrastructural double-staining method of rat growth hormone and its mRNA using LR White resin: a technical note. Histochem J 30:105-109

Matsuno A, Ohsugi Y, Utsunomiya H, Takekoshi S, Osamura RY, Watanabe K, Teramoto A (1994a) Ultrastructural distribution of growth hormone, prolactin mRNA in normal rat pituitary cells: A comparison between preembedding and postembedding methods. Histochemistry 102:265-270

Matsuno A, Ohsugi Y, Utsunomiya H, Takekoshi S, Sanno N, Osamura RY, Watanabe K, Teramoto A, Kirino T (1995) Changes in the ultrastructural distribution of prolactin and growth hormone mRNAs in pituitary cells of female rats after estrogen and bromocriptine treatment, studied using in situ hybridization with biotinylated oligonucleotide probes. Histochem Cell Biol 104:37-45

Matsuno A, Teramoto A, Takekoshi S, Utsunomiya H, Ohsugi Y, Kishikawa S, Osamura RY, Kirino T, Lloyd RV (1994b) Application of biotinylated oligonucleotide probes to the detection of pituitary hormone mRNA using Northern blot analysis, in situ hybridization at light and electron microscopic levels. Histochem J 26:771-777

Matsuno A, Utsunomiya H, Ohsugi Y, Takekoshi S, Sanno N, Osamura RY, Nagao K, Tamura A, Nagashima T (1996) Simultaneous ultrastructural identification of growth hormone and its messenger ribonucleic acid using combined immunohistochemistry and non-radioisotopic in situ hybridization: a technical note. Histochem J 28:703-707

Morel G, Chabot JG, Gossard F, Heisler S (1989a) Is atrial natriuretic peptide synthesized and internalized by gonadotrophs? Endocrinology 124:1703-1710

Morel G, Dihl F, Gossard F (1989b) Ultrastructural distribution of growth hormone (GH) mRNA and GH intron 1 sequences in rat pituitary gland: effects of GH releasing factor and somatostatin. Mol Cell Endocrinol 65:81-90

Morey AL, Ferguson DJP, Fleming KA (1995) Combined immunocytochemistry and non-isotopic in situ hybridization for the ultrastructural investigation of human parvovirus B19 infection. Histochem J 27:46-53

Pomeroy ME, Lawrence JB, Singer RH, Billings-Gagliardi S (1991) Distribution of myosin heavy chain mRNA in embryonic muscle tissue visualized by ultrastructural in situ hybridization. Dev Biol 143:58-67

Singer RH, Langevin GL, Lawrence JB (1989) Ultrastructural visualization of cytoskeletal mRNAs and their associated proteins using double-label in situ hybridization. J Cell Biol 108:2343-2353

Trembleau A, Calas A, Fevre-Montange M (1990) Ultrastructural localization of oxytocin mRNA in the rat hypothalamus by in situ hybridization using a synthetic oligonucleotide. Brain Res Mol Brain Res 8:37-45

Wolber RA, Beals TF, Maassab HF (1989) Ultrastructural localization of Herpes simplex virus RNA by in situ hybridization. J Histochem Cytochem 37:97-104

Chapter 16

Whole Mount in Situ Hybridization for mRNA Detection in Chick Embryos

HIDEYO OHUCHI, AND SUMIHARE NOJI

Introduction

Analysis of the spatial expression patterns of genes in animal embryos is essential to understanding functions of the genes especially for developmental biology. In situ hybridization with whole embryos is a powerful method to observe overall expression patterns of genes in embryos. A method for the whole mount in situ hybridization was developed after non-radioactive probes, especially digoxigenin-labeled RNA probes, became commercially available. Furthermore, using two different non-radioactive probes, expression patterns of two different genes can be observed simultaneously in an embryo, as a pattern of two colors. Sections of the stained embryos can be used to identify cells expressing the mRNA in the embryo. Thus, this method allows the identification of mRNA within intact embryos at both regional and cellular levels.

Outline

The entire protocol of a whole-mount in situ hybridization is illustrated in Fig. 1 (Noji et al. 1997).

Hideyo Ohuchi, The University of Tokushima, Department of Biological Science and Technology, Faculty of Engineering, 2-1 Minami-Jyosanjima-cho, Tokushima City, 770, Japan; *Correspondence to* Sumihare Noji, The University of Tokushima, Department of Biological Science and Technology, Faculty of Engineering, 2-1 Minami-Jyosanjima-cho, Tokushima City, 770, Japan (*phone* +81-886-56-7528; *fax* +81-886-56-9074; *e-mail* noji@bio.tokushima-u.ac.jp)

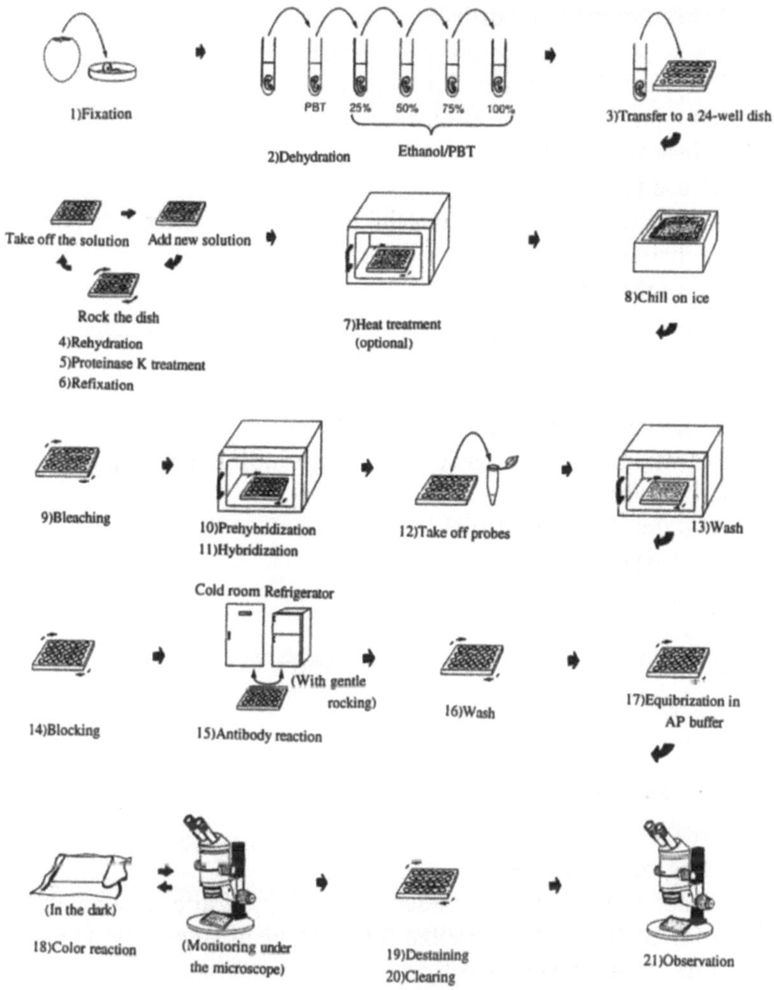

Fig. 1. Protocol of a whole-mount in situ hybridization

Materials

Equipment

- Dissection microscope (e.g., Leica MZ series, Zeiss Stemi SV11)
- Agarose gel electrophoresis apparatus (RNase-free)
- Variable speed shaking platform (e.g., Belly Dancer)
- Heater blocks that can accommodate 1.5-2 ml microtubes
- Hybridization oven

Labware

- 24-well disposable dish

Reagents

- Enzymes (Boehringer, Promega etc.)
- RNA polymerase (appropriate polymerase; T7, T3, SP6)
- RNase Inhibitor
- DNase (RNase-free)
- Nucleotide mix (pH7.5; or DIG RNA Labeling Mix, 10x conc; Boehringer)
  - 10mM GTP
  - 10mM ATP
  - 10mM CTP
  - 6.5mM UTP
  - 3.5mM digoxigenin-UTP
- 5x transcription buffer (or use buffer supplied by supplier of RNA polymerase)
  - 200mM Tris HCl (pH7.9)
  - 30mM $MgCl_2$
  - 10mM spermidine
  - 50mM NaCl
- Reagents for agarose gel electrophoresis (RNase-free):
  - Agarose
  - 1x TAE or 1x TBE
  - 10mg/ml Ethidium bromide
  - DNA or RNA marker
- 4M LiCl
- 10mg/ml yeast tRNA

  Yeast tRNA powder (molecular biology grade) is available from GIBCO-BRL. Dissolve the powder in sterile dH_2O and store in aliquats at -20°C.
- Glacial acetic acid
- Ethanol (200-proof)
- dH_2O (RNase-free)
- Buffers for alkaline hydrolysis (store in aliquots at -20°C)
  - 80mM $NaHCO_3$
  - 120mM Na_2CO_3
- PBS (phosphate buffered saline; autoclave and store at 4°C)
  - 10mM Sodium Phosphate (pH 7.4)
  - 100mM NaCl
  - Autoclave.
- 0.2M Sodium phosphate; solution A: solution B= 22.6: 77.4

  - Solution A: 0.2M NaH_2PO_4 (Store at room temperature (RT)
  - Solution B: 0.2M Na_2HPO_4 (Store at RT)
- PBT (store at RT)
  - 0.1% Tween-20 in PBS

Note: Do not autoclave after adding Tween-20.

- 4% (w/v) Paraformaldehyde in PBS (PFA; freshly prepared)
- 20% PFA stock in PBS (store at 4°C).

Note: Use an electron microscopy grade PFA. Heat at 65°C and add a few drops of 5N NaOH solution with occasional agitation until dissolved, and cool on ice. It should be stored at 4°C in tightly closed bottles and discarded after a few weeks.

- Ethanol series (store at RT)
  - 25% ethanol in PBT
  - 50% ethanol in PBT
  - 75% ethanol in PBT

Note: Do not use when precipitates appear.

  - Ethanol (190-proof)

- 2-20μg/ml Proteinase K (freshly prepared),
  - 20mg/ml stock in sterile dH_2O

Note: Proteinase K powder is available from Boehringer. After dissolving in dH_2O, store in aliquots at -20°C.

- 2mg/ml Glycine/PBT (freshly prepared; optional)
- 0.2% Glutaraldehyde (GTA), 4% PFA/PBT (freshly prepared)
  - 20% GTA stock solution (electron microscopy fixative grade; store at 4°C)
  - 20% PFA stock solution (store at 4°C)
- 6% H_2O_2/PBT (freshly prepared; optional)
  - 30% H_2O_2 (store at 4°C)
- Prehybridization solution (store at -20°C)
  - 50% formamide
  - 5x SSC (pH4.5)
  - 1% SDS
  - 50μg/ml yeast tRNA
  - 50μg/ml heparin
- 20x SSC (pH 4.5)

- 3M NaCl
- 0.3M sodium citrate

Note: Adjust pH with citric acid

- Solution I (store at RT)
  - 50% formamide
  - 5x SSC (pH4.5)
  - 1% SDS
- Solution III (store at RT)
  - 50% formamide
  - 2x SSC (pH4.5)
  - 1% SDS
- TBST (store at RT)
  - 100mM Tris HCl (pH 7.5)
  - 150mM NaCl
  - 0.1 or 1% Tween-20
- 1.5% Blocking reagent (Boehringer)

Note: Dissolve at 65°C in TBST. Store in aliquots at -20°C.

- Embryo powder (optional): Preparation of embryo powder
  - 1. Wash appropriate stage embryos in ice-cold PBS.
  - 2. Homogenize the embryos in fresh ice-cold PBS.
  - 3. Add 4 vol. of ice-cold acetone, mix, and incubate at RT for 30 mins.
  - 4. Centrifuge at 10000 g for 10 mins, remove the supernatant, and then wash the pellet with ice-cold acetone and spin again.
  - 5. Spread the pellet out on a sheet of filter paper.
  - 6. Dry the pellet in a disiccator overnight.
  - 7. Grind the dried pellet into a fine powder and store in an air-tight tube at 4°C.
- Anti-DIG-AP (alkaline phosphatase) conjugate, Fab fragment (Boehringer; store at 4°C)

Note: Do not freeze.

- NTMT (freshly prepared)
  - 100mM Tris HCl (pH 9.5)
  - 100mM NaCl
  - 50mM $MgCl_2$
  - 1% Tween-20
- AP buffer (freshly prepared)

- 100mM Tris HCl (pH 9.5)
- 100mM NaCl
- 50mM $MgCl_2$
- 1% Tween-20
- 5% polyvinyl alcohol (PVA; 70-100 kDa; Sigma)

Note: Dissolve PVA in a microwave oven. After cooling down at RT, add $MgCl_2$ and Tween-20, and mix up to the final volume with dH_2O.

- Nitroblue tetrazolium chloride (NBT) stock solution (store in dark at -20°C)
  - 75mg/ml NBT (Boehringer) in 70% dimethylformamide (DMF)
- X-phosphate (BCIP) stock solution (store in dark at -20°C)
  - 50mg/ml BCIP (Boehringer) in DMF
- Color reaction solution (freshly prepared)
  - 4.5µl 75mg/ml NBT
  - 3.5µl 50mg/ml BCIP
  - 1ml AP buffer
  - Filter
- 25%, 50% Glycerol/PBT

Procedure

Preparation of labeled RNA probe

1. Mix these reagents in the following order at RT:

dH_2O (RNase-free)	8µl
5 x transcription buffer	4µl
0.1M dithiothreitol (DTT)[1]	2µl
DIG RNA Labeling Mix, 10 x conc.	2µl
Linearized template DNA[2] (1g/l)	1µl (1µg)
RNase inhibitor	1µl (20-40U)
RNA polymerase	2µl (20-40U)
Total volume is	20µl

Note: When DTT is included in the transcription buffer, add the same volume of dH_2O.

Note: It is not advisable to cut the plasmid DNA with the restriction enzymes that create 3'-protruding ends, such as *Apa I, Kpn I, Pst I, Sac I,* and *Sph I.*

2. Incubate at 37°C for 1hr.

3. Add 2μl RNA polymerase, 1μl RNase inhibitor. Incubate at 37°C for 1hr.

4. Chill on ice. Remove a 1μl aliquot and run on an agarose gel (1-2%) to estimate the amount synthesized.

Note: This gel must be RNase-free, but does not need to be denaturing. An RNA band 10-fold more intense than the plasmid band should be seen, indicating that 10μg probe has been synthesized. Use a nucleotide size marker and estimate the size of the riboprobe synthesized. Unexpected larger size RNA could give some backgrounds.

5. Add 1μl RNase inhibitor, 2μl DNase (RNase-free; Promega, 1U/μl). Incubate at 37°C for 15 mins.

6. Add 3μl 4M LiCl, 1μl yeast tRNA (10mg/ml), 100μl ethanol (200-proof). Mix and incubate at -20°C for >1hr.

7. Spin in a microfuge for 20 mins at 4°C. Wash pellet with 80 % ethanol and air-dry the pellet.

8. (Optional; If the transcript is greater than 0.5 kb in length, reduce the average size to 500 bases). Add an equal volume (100μl each) of 80 mM $NaHCO_3$, 120 mM Na_2CO_3 and heat at 60°C for a period of time given by: time (t; min) =(L-Lf)/(k x L x Lf), where L is the starting length of the transcript (kb), Lf is the the desired length (kb), and k is 0.11 (cuts/kb/min). When Lf is 0.5, t = (L-0.5)/(0.055L).

9. Chill on ice. Add 1μl glacial acetic acid on ice, mix, add 20μl 3M sodium acetate, 560μl ethanol. Mix and incubate at -20°C for ›1hr.

10. Spin in a microfuge for 20 mins at 4°C. Wash pellet with 80% ethanol, and air-dry the pellet.

11. Redissolve in prehybridization buffer at 0.1 - 0.5 µg/µl. Store at -80°C. Use appropriate volume for hybridization mix at 100-500 ng/ml.

Note: We have found it better to use long probes as they are much more sensitive than short probes.

Fixation and dehydration of embryos

1. Dissect embryos in PBS, removing as much extraneous tissue as possible. To avoid the trapping of reagents, open any cavities, such as the brain vesicle, optic vesicles, otic vesicles, and heart, using a needle (27 gauge or less).
2. Fix in paraformaldehyde fixative at 4°C, 2 hrs to overnight.
3. Rinse the embryos with PBT, twice for 5 mins.
4. Dehydrate the embryos by washing for 5 mins each in a graded ethanol series diluted in PBT (25% ethanol, 50% ethanol, 75% ethanol) and then twice with 100% ethanol (190-proof). The embryos can now be stored at -20°C for at least several months.

Pretreatment and hybridization of the embryos

Note: Unless otherwise stated all treatment are done at RT.

Leave a little liquid above the embryos after each wash or you will damage the embryos.

All washes are carried out with gentle rocking.

1. Transfer the embryos into a 24-well dish. Rehydrate by taking through the ethanol/ PBT in reverse and then wash twice with PBT.
2. Treat the embryos with 2 20µg/ml proteinase K in PBT for up to 15 mins

Note: The length and time of treatment depends on the experimental purpose (e.g., the size of tissue, the stage of embryos, epithelial or mesenchymal tissue) and should be optimized. We usually treat chick embryos with 10µg/ml for X mins (X=

the number of Hamburger-Hamilton's stage of the embryos examined). Try at 37°C when stronger treatment is needed on larger and mesenchymal tissues.

3. Wash for 5 mins with 2mg/ml glycine in PBT.

Note: This step is for complete inactivation of proteinase K and can be omitted.

4. Wash twice for 5 mins each with PBT.
5. Refix the embryos with fresh 0.2% glutaraldehyde/4% paraformaldehyde in PBT for 20 mins.
6. Rinse the embryos twice for 5 mins with PBT.
   (Optional) Incubate the embryos in PBT at 70 °C for 40 to 75 mins. Chill on ice.

Note: This step (heat treatment) is for inactivation of endogenous peroxidase and can be omitted in the case of early embryos.

7. (Optional; for embryos older than about 2 days) Bleach for 1hr in 6% H_2O_2/PBT. Wash with PBT, three times for 5 mins.
8. Remove most of the liquid, add 1 ml prewarmed prehybridization solution and allow the embryos to sink.
9. Replace with fresh 1 ml prehybridization solution and incubate at 65 °C in hybridization oven for more than 1h. The embryos can be stored at -20°C at this point.
10. Add 700μl hybridization mix and incubate at 65°C overnight with rocking.

Hybridization mix

Prehybridization buffer	600μl
Diluted probe (0.7-3.5 ng/μl)	100μl

Note: Cover the 24-well dish with Saran Wrap and put a water tray in the oven to avoid drying up.

Posthybridization washing and antibody reaction

Note: All washes are carried out with sufficient shaking to agitate the embryos, otherwise high backgrounds result.

We usually use 1ml solution in a well of the 24-well dish.

After each 65°C wash, put the 24-well dish on a heater block at 65°C, then change supernatant individually so samples don't cool. Keep solutions I and III at 65°C in hybridization oven.

1. Wash with solution I, three times for 20 mins at 65°C.

Note: Diluted probe can be stored at -20°C and reused.

2. Wash with solution III, twice for 5 mins at 65°C.
3. Wash with solution III, three times for 20 mins at 65 °C.
4. Rinse with TBST, twice for 10 mins at RT.
5. Preblock the embryos with 1.5% blocking reagent in TBST for 60-90 mins.
6. During this, preabsorb the antibody for 1hr with embryo powder as described below.
   1) Weigh out 3mg embryo powder into a microtube, add 0.5ml TBST and heat at 70°C for 30 mins.
   2) Cool on ice and add 5μl 1.5% blocking reagent and 1μl AP-anti-DIG-antibody.
   3) Shake gently at 4°C for 1hr, then spin in a microfuge at 4°C for 10 mins. Recover the supernatant and dilute it to 2 ml with 1.5% blocking reagent.
7. Incubate with the diluted AP-anti-DIG antibody and rock overnight at 4°C.

Post antibody reaction wash and color reaction

1. Wash with TBST, four times for 5 mins.

Note: Use TBST containing 1% Tween-20 in this step. The higher concentration of Tween-20 helps to give low backgrounds.

2. Wash with TBST, five to seven times for 1hr.

Note: The embryos can be stored at 4 °C at this point. The overnight wash may give lower backgrounds.

3. Wash twice with NTMT for 5 mins.
4. Wash twice with 1:1 NTMT: AP buffer for 10 mins.
5. Wash six times with AP buffer for 20 mins.
6. Incubate in the dark with color reaction solution (NBT/BCIP/AP buffer) with gentle rocking for the first 5 mins. Periodically monitor the reaction after the first 30 mins.

Note: It is not advisable to examine too often under the microscope because the lights will oxidize the substrate and may give backgrounds.

7. When a strong signal is produced and/or excessive background is observed, stop the color reaction by washing several times with NTMT.
8. Wash 5 times in PBT for 5 mins. Stained embryos can be stored in PBT or in 50% glycerol/PBT at 4°C.

Note: The color reaction product is converted to a blue color by washing in PBT. The embryos can be cleared in 50% glycerol/PBT.

The simultaneous detection of different target mRNAs can be done by using probes labeled with different haptens (e.g., anti-fluorescein) and different substrates for color reaction (e.g., Fast Red; Dietrich et al. 1997).

Destaining

If the embryos are overstained, it is possible to partially destain them in ethanol, but care has to be taken in case ethanol removes too much signal or by brief incubation in ethanol.

1. Incubate in PBT for 1hr at RT.
2. Wash with 50% ethanol in PBT for 5 mins.
3. Wash with 100% ethanol (190-proof) for up to 2hrs.
4. Wash with 50% ethanol in PBT for 5 mins.
5. Wash with PBT for 5 mins.

Histological analysis of stained embryos

To process for histological analysis, the stained embryos can be sectioned with vibratome after embedding in Gellan Gum or agarose (Osumi et al. 1997), or with microtome after paraffin-embedding according to Sasaki and Hogan (1993).

Results

Tipical results are shown in Fig. 2.

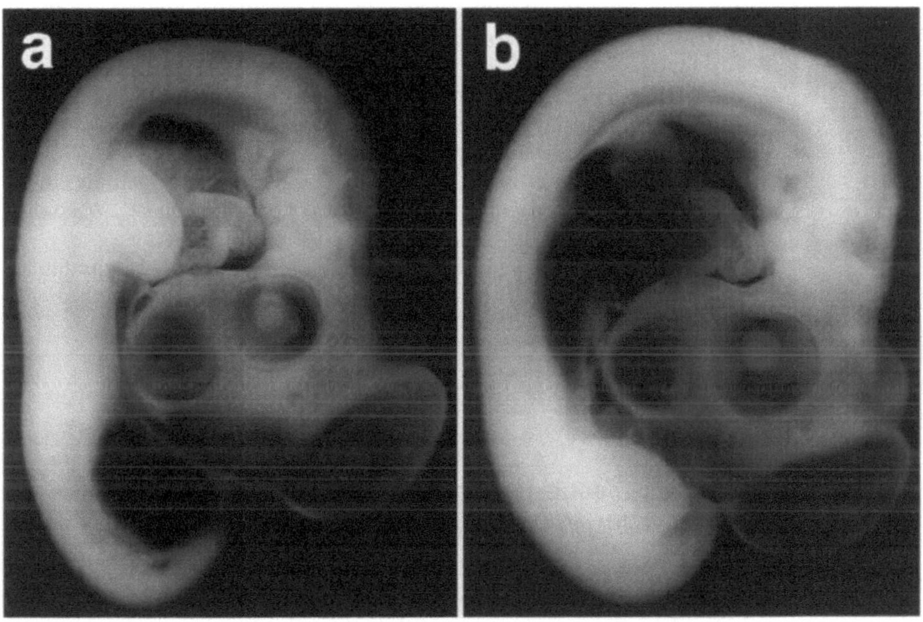

Fig. 2. Whole-mount in situ hybridization of chick embryo. (a, b) Four-day chick embryos hybridized with probes for the T-box genes *Tbx4* (a) and *Tbx5* (b) by the method described in the text. *Tbx5* gene expression is observed in the wing bud, whereas *Tbx4* is expressed in the leg bud.

References

Block MD, Debrouwer D (1993). RNA-RNA in Situ Hybridization using Digoxigenin-Labeled Probes: The use of High-Molecular-Weight Polyvinyl Alcohol in the Alkaline Phosphatase Indoxyl-Nitroblue Tetrazolium Reaction. Analytical Biochemistry 215: 86-89

Dietrich S, Schubert F, Lumsden A (1997). Control of dorsoventral pattern in the chick paraxial mesoderm. Development 124: 3895-3908

Noji S. (1997) A practical protocol for whole-mount in situ hybridization. In: Immunostaining and In situ hybridization (Yodosha The Protocol Series), pp80-93. Tokyo: Yodosha

Osumi N, Hirota A, Ohuchi H, Nakafuku M, Iimura T, Kuratani S, Fujiwara M, Noji S, Eto K (1997). Pax-6 is involved in the specification of hindbrain motor neuron subtype. Development 124: 2961-2972

Riddle RD, Johnson RL, Laufer E., Tabin C.(1993). Sonic hedgehog mediates the polarizing activity of the ZPA. Cell 75: 1401-1416

Rosen B, Beddington SP (1993).Whole-mount in situ hybridization in the mouse embryo: gene expression in three dimensions. Trends Genet 9: 162-167

Sasaki H, Hogan BLM (1993). Differential expression of multiple fork head-related genes during gastrulation and axial pattern formation in the mouse embryo. Development 118: 47-59

Wilkinson DG (1992). Whole mount in situ hybridization of vertebrate embryos. In: In Situ Hybridization, A Practical Approach (ed.D.G.Wilkinson),pp.75-83. Oxford:IRL press at Oxford University Press

Wilkinson DG, Nieto MA (1993). Detection of Messenger RNA by in Situ Hybridization to Tissue Sections and Whole Mounts. Methods Enzymol 225: 361-373

Suppliers

Boehringer Mannheim Biochemicals (Roche Molecular Biochemicals), http://biochem.boehringer.com

GIBCO BRL (Life Technologies), Gaithersburg MD, USA (*phone* +1-301-840-8000, http://www.lifetech.com

Promega Corporation, 2800 Woods Hollow Road, Madison WI, 53711-5399, USA (*phone* +1-608-274-4330; *fax* +1-608-277-2516, http: //www.promega.com

Part III

Chapter 17

Introduction to the Detection of Transcription Regulatory Proteins

TAKEHIKO KOJI

Does an increase in the level of a certain mRNA mean the increase in transcriptional activity?

We have already described how to localize specific mRNA and DNA in cell or tissue preparations in the previous section. As in the case of northern blot analysis of mRNA expression, it is apparent that the information on the expression of mRNA by in situ hybridization in itself represents only a steady-state level of gene expression. Consequently, we cannot know whether an elevated level of certain mRNA is due to an increased rate of transcription of the gene or due to a decreased rate of degradation of the mRNA. Thus, for a more precise understanding of the regulation of gene expression, one would require an analysis of the expression of trans-acting transcription regulatory factors at the individual cell level. Those proteins, such as c-Myc (Kerkhoff et al. 1991), c-Jun (Curran and Franza 1988), CREB (Meyer and Habener 1993) and various steroid hormone receptors (Beato 1989), bind specific consensus sequences in nuclear DNA known as responsive element and regulate transcriptional activity of the genes with the element. To date, however, in most of the transcription regulatory factors reported so far, there is only a little information on their cellular distribution because of the limited availability of specific antibodies which allow us to perform immunohistochemical demonstration of those proteins. Considering that the expression level of those proteins is generally too low to be isolated at a sufficient amount to raise antibodies against them, we needed to explore a new methodology.

Correspondence to Takehiko Koji, Nagasaki University, Dept. of Histology and Cell Biology, 1-12-4, Sakamoto, Nagasaki, 852-8523, Japan (*phone* +81-095-849-7025; *fax* +81-095-849-7028; *e-mail* tkoji@net.nagasaki-u.ac.jp)

Transcription regulatory factors

As shown in Fig. 1, transcription regulatory factors, which regulate transcriptional activity of a gene, are usually divided into two categories; one consists of general transcription factors which include RNA polymerase II and various factors and bind to TATA box as a transcription machinery complex and regulate the transcription in most of the genes, and the other consists of gene specific transcription regulatory factors which bind to a consensus sequence in a specific responsive element and regulate the transcription of the gene specifically. What we are interested in now is the latter factors.

Those proteins can be firstly identified as factors bound to a specific DNA element which is involved in the regulation of the gene transcription, so the base sequence of the binding site is determined at the same time by DNase I foot-printing. Therefore, in many cases we can know the specific base sequences of responsive element to be recognized and bound by the regulatory protein, even though no other information on the protein is available.

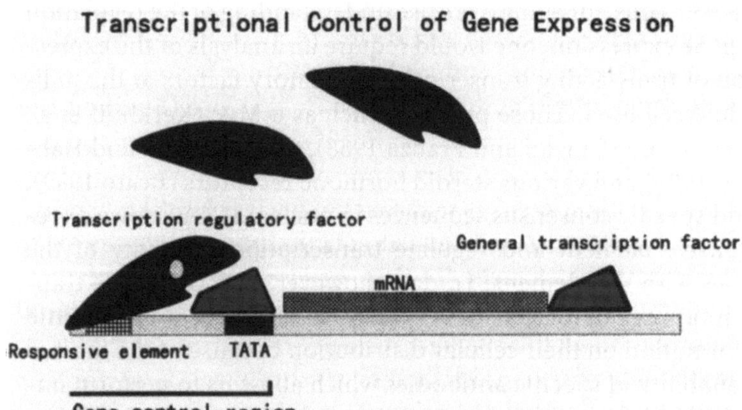

Fig. 1. Schematic presentation of the regulation of gene transcription. There are two categories in transcription regulatory proteins. One is a group of proteins forming transcriptional machinery and is involved in the transcription of genes directly by binding to TATA box as general transcription regulatory factors. Another group consists of gene specific transcription regulatory factors, which bind to the specific responsive element consensus sequence in the gene regulatory region of the target genes and are activated to control the activity of transcriptional machinery.

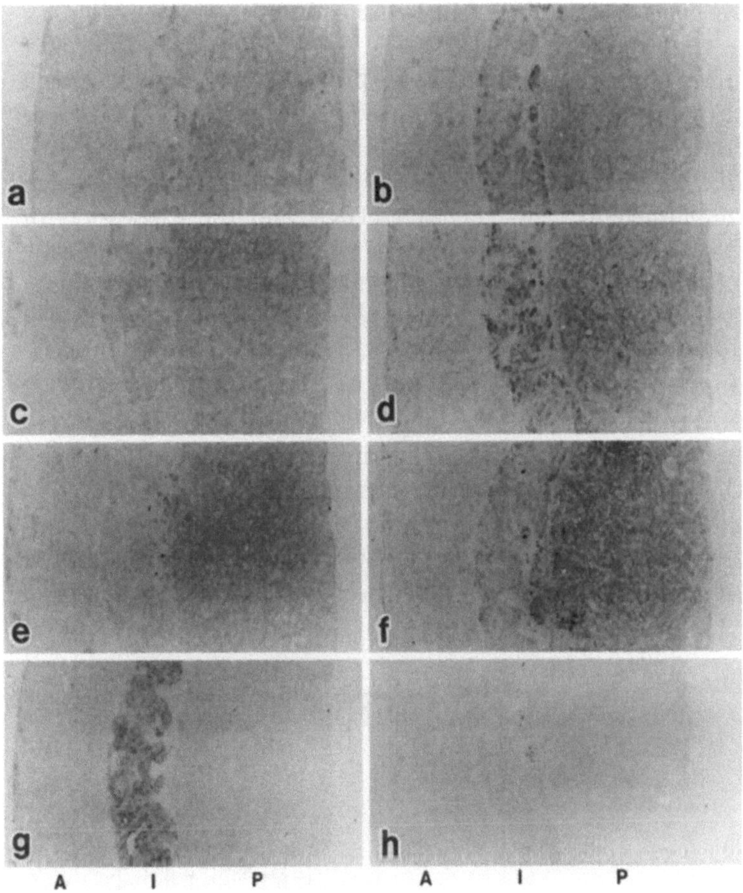

Fig. 2. In situ localization of POMC mRNA in the paraffin sections of rat pituitary gland. As described in the text, the paraffin sections of rat pituitary gland were hybridized in situ with DNP-labeled POMC cDNA (a, c, e, g) and DNP-labeled pBR 322 DNA (b, d, f, h) and the signal was detected enzyme-immunohistochemically (Koji et al. 1988). In this experiment, the effects of increasing concentrations of proteinase K (37 °C, 15 mins) on the signal detection was investigated, since the optimization of proteinase digestion conditions is the most important factor in achieving the best performance of nonradioactive in situ hybridization (c.f., Chapter 15). (a, b); 0 μg/ml, (c, d); 1 μg/ml, (e, f); 10 μg/ml, (g, h); 100 μg/ml. A; anterior pituitary, I; intermediate pituitary, P; posterior pituitary. At 100 μg/ml of proteinase K, the specific signal for POMC mRNA was observed in the intermediate pituitary cells and some cells of anterior pituitary. However, a clear staining with the similar distribution pattern to that of POMC mRNA was obtained with pBR 322 DNA at lower concentrations of proteinase K.

Development of Southwestern histochemistry

During our efforts to optimize the protocol for nonradioactive in situ hybridization using rat pituitary sections in 1985, we encountered a strange phenomenon. In an experiment, we hybridized the pituitary sections with dinitrophenyl (DNP)-labeled proopiomelanocortin (POMC) cDNA and the expression of POMC mRNA was analyzed enzyme-immunohistochemically with anti-DNP antibody. As shown in Fig. 2, the specific staining of POMC mRNA was found in intermediate pituitary cells as well as some cells of anterior pituitary at a relatively high concentration of proteinase K. Surprisingly, a bacterial plasmid pBR 322 DNA, which was used as a negative control probe, gave rise to some staining in a very similar distribution pattern to that of POMC mRNA, but at low concentrations of proteinase K. Later, we realized that the distribution of the "specific" protease- sensitive staining with pBR 322 DNA was quite similar to that of glucocorticoid receptor (GR). Therefore, it was really exciting

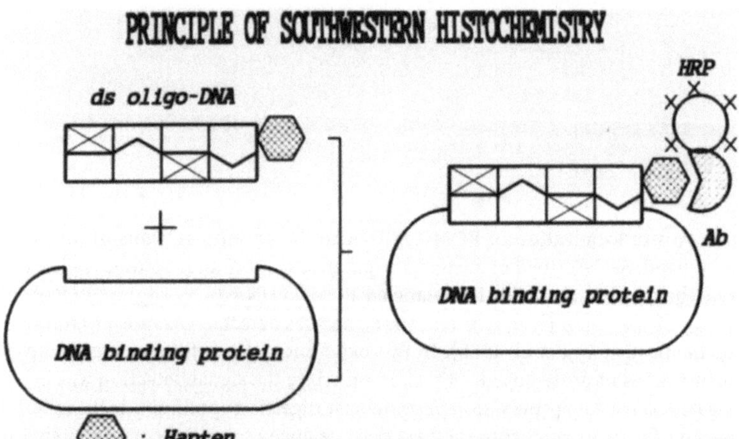

Fig. 3. Schematic presentation of Southwestern histochemistry. (1) The (+) and (-) strands of specific consensus sequence DNA in target response element are synthesized and annealed after synthesis. The 3' ends of both strands are labeled with a haptenic compound such as thymine-thymine (T-T) dimer or digoxigenin. (2) The double-stranded oligo-DNA with a hapten is reacted with tissue sections. (3) Finally, horseradish peroxidase (HRP)-labeled anti-hapten antibody is reacted with the sections and the sites of HRP are visualized with DAB and hydrogen peroxide in the presence of nickel and cobalt ions (see Chapter 15).

to find the report in which pBR 322 DNA contains glucocorticoid responsive element (GRE) consensus sequences at five sites and actually can bind to GR on southwestern blots (Tully and Cidlowski 1987). Our previous finding now can be interpreted in a way that pBR 322 DNA may bind to GR in the tissue section. This interpretation is also consistent with the fact that GR consists of three autonomous domains; trans-activation domain, DNA-binding domain and hormone-binding domain, and regulates gene expression through its binding to specific DNA sequences, called "GRE", located in the vicinity of the regulated gene (Evans 1988; Beato 1989).

Based upon the accumulated data, we attempted to localize GR by utilizing the specific interaction between GR and GRE (later the method was named Southwestern histochemistry (Koji et al 1994) owing to the apparent similarity in principle to that of southwestern analysis (Miskimins et al. 1985)); frozen sections of rat liver were reacted with thymidine-thymidine (T-T) dimerized pBR 322 DNA and then the T-T dimers were detected enzyme-immunohistochemically by the successive use of rabbit anti- (T-T dimer) antibody and HRP-anti-(rabbit IgG). Consequently, specific staining was confirmed in the nuclei, nuclear membranes as well as cytoplasm of hepatocytes (Koji et al. 1990, 1992). The success encouraged us to combine the method with the use of synthetic double-stranded oligo-DNA, as described schematically in Fig. 3. In principle, the use of haptenic oligo-DNA is more specific and more convenient than segments of DNA (Koji et al. 1992) or radiolabeled oligo-DNA.

As a model system to demonstrate the usefulness of haptenized oligo-DNA for the localization of specific transcription factors at the level of individual cells, cyclic adenosine 3',5'-monophosphate (cAMP) responsive element (CRE) binding protein was localized in the frozen sections of rat brain and small intestine (Fig. 4) (Koji et al. 1994). We synthesized the CRE consensus base sequence (TGACGTCA) together with three TTA repeats at 5'-end and allowed it to be annealed, as given below:

```
** ** **
TTATTATTA-TGACGTCA
          ACTGCAGT-ATTATTATT
                   ** ** **
```

(**) = possible sites of T-T dimers

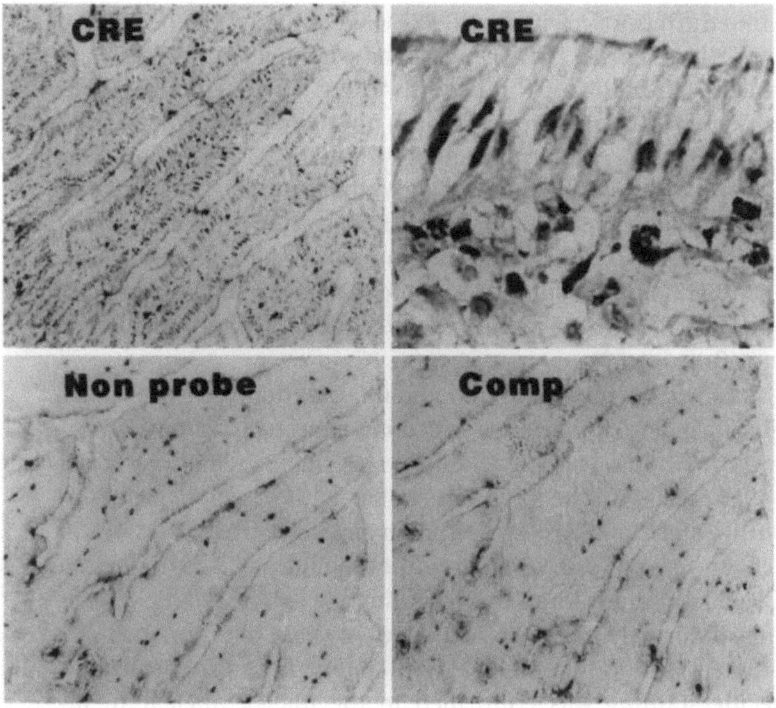

Fig. 4. In situ detection of CRE-binding proteins by Southwestern histochemistry in the fresh frozen sections of rat small intestine (Koji et al. 1994). The fresh frozen sections of rat small intestine were fixed with 4% paraformaldehyde in PBS and reacted with T-T dimerized CRE double-stranded oligo-DNA probe (CRE); the left panel is a low-power view and the right panel is a high-power view. (Non probe); a section was processed omitting only the probe DNA. (Comp); a section was reacted with the CRE probe in the presence of an excess amount of non-labeled CRE DNA (102-fold). In CRE panels, the strong staining of epithelial nuclei of small intestine was found, whereas in a Non probe section and a Comp section no signal was detected.

Before use, the double-stranded CRE oligo-DNA was T-T dimerized by UV irradiation. When the sections of those tissues, which were known to be responsive to cAMP, were reacted with the haptenized probe, the brain and intestinal epithelial nuclei were heavily stained and the staining intensities could be altered by physiological manipulation (Koji et al. 1994).

Although the list of transcription factors and their responsive elements is rapidly growing, antibodies are not available for most of these newly identified proteins. This method should be useful for the localization of those factors at an individual cell level and for a better understanding of transcriptional regulation of specific mRNA expression. Also, application of Southwestern histochemistry to EM will be appropriate to explore the relationship between transcription factors and chromatin structure in nuclci.

References

Beato M (1993) Gene regulation by steroid hormones. Cell 56:335-344

Curran T, Franza RB (1988) Fos and Jun: The AP-1 connection. Cell 55: 395-397

Evans RM (1988) The steroid and thyroid hormone receptor. Science 240: 889-895

Kerkhoff E, Bister K, Klempnauer K-H (1991) Sequence-specific DNA binding by Myc proteins. Proc Natl Acad Sci USA 88:4323-4327

Koji T, Komuta K, Nozawa M, Yamada S, Nakane PK (1994) Localization of cyclic adenosine 3',5'-monophosphate-responsive element (CRE)-binding proteins by southwestern histochemistry. J Histochem Cytochem 42:1399-1405

Koji T, Moriuchi T, Nakane PK (1988) Improved tissue preparation for in situ localization of specific mRNA using non-radioactive DNA probes: Effects of protease digestion and probe size on signal detection in frozen and paraffin sections of rat pituitary glands. Acta Histochem Cytochem 21:187-200

Koji T, Yamada S, Izumi S, Nakane PK (1990) Oligo-histochemistry. A new approach to localize DNA-binding proteins. J Histochem Cytochem 38:1052

Koji T, Yamada S, Kayashima K, Nakane PK (1992) A new approach to localize glucocorticoid receptor using DNA probe containing glucocorticoid responsive element DNA consensus sequences. Acta Histochem Cytochem 25:681-687

Meyer TE, Habener JF (1993) Cyclic adenosine 3',5'-monophosphate response element binding protein (CREB) and related transcription- activating deoxyribonucleic acid-binding proteins. Endocrine Rev 14:269-290

Miskimins WK, Roberts M, McClelland A, Ruddle FH (1985) Use of a protein-blotting procedure and a specific DNA probe to identify nuclear proteins that recognize the promoter region of the transferrin receptor gene. Proc Natl Acad Sci USA 82:6741-6744

Tully DB, Cidlowski JA (1987) pBR322 contains glucocorticoid regulatory element DNA consensus sequences. Biochem Biophys Res Commun 144:1-10

Chapter 18

Southwestern Histochemistry

KO KOMUTA, TAKASHI KANEMATSU AND TAKEHIKO KOJI

Introduction

Abnormal expression of oncogenes or proto-oncogenes are frequently related with cancer development. The epidermal growth factor receptor (EGFR) gene has been found to be amplified and/or expressed at high levels in various types of tumor cells including A431 human epidermoid carcinoma cells [1,2,3]. Although the over-production of EGFR protein generally correlates with an elevated level of its mRNA, it does not always accompany EGFR gene amplification [4,5].Therefore, to understand the abnormal nature of cancer growth, the analysis of their gene expression at the transcriptional level as well as the transcript level is required.

The method, named southwestern histochemistry, may enable us to analyze the expression of transcription regulatory factors as specific DNA binding proteins at individual cell level.

Principle of Southwestern histochemistry

The principle of Southwestern histochemistry is almost the same as with Southwestern analysis. We synthesized the set of (+) and (-) oligo-DNAs consisted of specific resposive element on an automated DNA synthesizer . These synthesized oligo-DNAs were purified and labeled. The (+) and (-) strands of each synthesized oligo-DNAs were mixed, denatured and annealed. We re-

---

Correspondence to Ko Komuta, Takashi Kanematsu, Nagasaki University School of Medicine, Department of Surgery II, 1-12-4, Sakamoto, Nagasaki, 852, Japan; Takehiko Koji, Nagasaki University School of Medicine, Department of Histology and Cell Biology, 1-12-4, Sakamoto, Nagasaki, 852, Japan

acted haptenized double-stranded DNAs with a section and examined a signal immunochistochemically.

Procedure

Practice of Southwestern histochemistry

Room

The contamination of RNase may be admitted. However, a contamination of DNase should be avoided. Actually, we experiment in the same room as for an immunohistochemical experiment.

However, a solution is treated by autoclave and sterilized glasses are used.

Synthesis of oligodeoxynucleotide (oligo-DNA)

We synthesized the set of (+) and (-) oligo-DNAs consisting of specific responsive element on an automated DNA synthesizer (Applied Biosystems, Model 391 PCR-Mate) . These synthesized oligo-DNAs were purified by an OPC column.

Haptenization of oligo-DNA

(a) Thymine-thymine (T-T) dimers
For haptenization with T-T dimers of the oligo-DNAs, we added 2 and 3 TTA repeats at 5',- and 3',- end of the selected sequences, respectively. The diluted (20-50 µsg/ml) oligo-DNA by UV irradiation (5000-10000 J/m^2) was haptenized. The optimal UV dose was determined by dot-blot hybridization [6].
(b) Dig-11-dUTP
The synthesized oligo-DNAs were labeled at 3'-end with Dig-11-dUTP by terminal deoxynucleotidyl transferase [7]. The haptenization was confirmed by dot-blot hybridization.

Annealing

Annealing of single-stranded(ss) oligo-DNAs to double stranded (ds) oligo-DNA. The (+) and (-) strands of each synthesized oligo-DNAs dissolved in 10 mM Tris-HCl buffer(pH7.4) containing 1 mM EDTA were denatured for 10 min at (Tm-25) °C and thereafter gradually cooled to 20 °C. Then the re-annealing confirmed that the oligo-DNAs were separated by PAGE with 4-20 % gradient gel and were stained with the silver staining kit (Daiichi Pure Chem. Co., Japan).

Southwestern blot analysis

Aliquotes of the nuclear extract in 62.5 mM Tris/HCl (pH6.8) containing 2% SDS,15% glycerol, and 0.0025% bromophenol blue were subjected to SDS-PAGE with a 4-20 % gradient gel in duplicate. One slab of the separated proteins was electrophoretically transferred onto nylon membranes according to the user manual of immunoblotting system (Daiichi Pure Chem. Co., Japan), and one of them was stained with the silver staining for visualization of protein bands. The molecular weight markers were as follows; bovine skeletal muscle myoglobin (17.2 kD), bovine erythrocyte carbonic anhydrase (30 kD), rabbit skeletal muscle aldolase (42.4 kD), bovine serum albumin (66.3 kD), E. coli -galactosidase (116.2 kD) and rabbit skeletal muscle myosin myosin (200 kD).

The blotted nylon membrane was processed according to Silva et al. [8] with a slight modification. Briefly, the blot was incubated with a preincubation solution containing 5 % nonfat dry milk dissolved in 10 mM Tris/HCl (pH 7.4) buffer containing 50 mM NaCl and 1 mM EDTA for overnight at 4 °C. Then the blot was reacted with 1 μg/ml Dig-labeled or T-T dimerized ds oligo-DNA dissolved in the preincubation solution for 3 hrs at RT. After successive washings with the preincubation solution (twice, 1 hr each) and PBS (3 times, 1 hr each), the blot was incubated for 1 hr in a blocking solution that contained 5 % BSA and 500 μg/ml normal sheep or goat IgG in PBS. The reaction with HRP-linked sheep anti-Dig IgG (Fab), which was diluted at 1:800 with the blocking solution, was performed for 1 hr. The T-T dimers were detected immunohistochemically using rabbit anti-(T-T dimer) IgG and HRP-Fab (anti-R IgG) in succession. As a chromogen solution of HRP, a mixture of DAB, H_2O_2, cobalt chloride and nickel ammonium sulfate was used [9].

To confirm the sequence specificity in the reaction of labeled ds oligo-DNA probes with the blotted proteins, the blot was reacted with 1 μg/ml probe in the presence of an excess amount (100 - 450-fold) of the unlabeled ds oligo-DNA.

Tissue preparation

In our past system, we used the fresh frozen sections alone. Immediately after resection, specimens (size;5-10 mm) were embedded in OCT compound and frozen under dry ice added ethanol.The frozen sections were stored under -80°C.

Southwestern histochemistry

The procedure of Southwestern histochemistry using Thymine-thymine (T-T) dimers is as follows:

1. Fresh frozen specimens: Immediately after resection, specimens (size;5-10 mm) were embedded in OCT compound and frozen under dry ice added ethanol. The frozen sections were stored under -80°C.
2. Dry up for fresh frozen section: (5-6μm) fresh frozen sections were dried up on gelatin coated slide glass.
3. Fixation: 4% PFA (10 mins 4°C)
4. PBS (RT, 5 min.x3)
5. Incubation: 50 mM Tris/HCl(pH7.4)/5% non-fat dry milk/ 50 mM NaCl/1 mM EDTA (RT, 1 hr)
6. Incubation: T-T dimerized double strand DNA(1.0-2.5μg/ml)/50 mM Tris/HCl(pH7.4)/5% non-fat dry milk/50 mM NaCl/1 mM EDTA (RT, overnight)
7. Washing: 50 mM Tris/HCl(pH7.4)/5% non-fat dry milk/50 mM NaCl/1 mM EDTA (RT,15 mins x2) PBS (RT,15mins x3)
8. Blocking: 500 μg/ml goat IgG/5% BSA/100 μg/ml salmon testis DNA/100 μg/ml yeast tRNA/NaN_3/PBS (RT,1 hr)
9. First antibody: Rabbit anti-T-T IgG / 5% BSA/100 μg/ml salmon testis DNA/100 μg/ml yeast tRNA/NaN_3/PBS (RT, overnight)
10. PBS (RT, 15 mins x4)
11. Second antibody: HRP-goat anti-rabbit IgG/5% BSA/100 μg/ml salmon testis DNA/100 μg/ml yeast tRNA/NaN_3/PBS (RT, 1 hr)
12. PBS (RT, 15 mins x4)
13. Visualization: 0.5 mg/ml DAB/0.025% $CoCl_2$/0.02% $NiSO_4(NH_4)_2SO_4$ /0.01% H_2O_2 /0.1M sodium phosphate buffer(pH7.2) (RT, 5 mins)

Procedure of Southwestern histochemistry (using digoxigenin (Dig) labeled ds oligonucleotides):

1. Fresh frozen specimens: Immediately after resection, specimens (size;5-10 mm) were embedded in OCT compound and frozen under dry ice added ethanol. The frozen sections were stored under -80°C.

2. Dry up for fresh frozen section: (5-6μm) fresh frozen sections were dried up on gelatin coated slide glass.
3. Fixation: 4% PFA (10 mins 4°C)
4. PBS (RT, 5 mins x3)
5. Incubation: 50 mM Tris/HCl(pH7.4)/5% non-fat dry milk/50 mM NaCl/1 mM EDTA (RT, 1 hr)
6. Incubation: Dig-ds oligo-DNA (0.25-2.0μg/ml)/50 mM Tris/HCl (pH7.4)/5% non-fat dry milk/50 mM NaCl/1 mM EDTA (RT, overnight)
7. Washing: 50 mM Tris/HCl (pH7.4)/5% non-fat dry milk/50 mM NaCl/1 mM EDTA (RT, 15mins x2) PBS (RT, 15mins x3)
8. Blocking: 500 μg/ml sheep IgG/5% BSA/100 μg/ml salmon testis DNA/100 μg/ml yeast tRNA/NaN_3/PBS (RT, 1 hr)
9. First antibody: HRP-sheep anti-Dig IgG(1:100)/5% BSA//100 μg/ml salmon testis DNA/100 μg/ml yeast tRNA/NaN_3/PBS (RT, overnight)
10. PBS (RT, 15 mins x4)
11. Visualization: 0.5 mg/ml DAB/0.025% $CoCl_2$/0.02% $NiSO_4(NH_4)_2SO_4$ /0.01% H_2O_2 /0.1M sodium phosphate buffer (pH7.2) (RT, 5 mins)

The important points are explained as follows:
(a) Fixation
At the beginning of the experiments, the effects of fixation on the reactivity in tissue sections were examined. When the sections were fixed in 4% PFA for 10 mins at 4°C , the stainings were strong. On the other hand, when the sections were fixed in acetone for 10 mins at 4°C , the stainings were very weak, and endogeneous peroxidase staining was remarkable. Consequently, all subsequent studies were carried out using 4% PFA fixation [10].
(b) Probe concentration
When the sections were reacted with the probe at various concentrations (0.5, 1.0, 2.5, or 5 μg/ml), the staining intensity was increased by increasing the probe concentration, reaching a plateu level at 1.0-2.5 μg/ml [10]. It is necessary to decide on the best probe concentration for each individual probe.

(c) Concentration of NaCl in the reaction mixture
It is known that an appropriate concentration of salt is required to keep the specific interaction between protein and DNA. Also, DNA binding proteins including transcription factors such as steroid hormone receptors and cyclic AMP responsive element binding proteins are extracted with a high-salt concentration buffer [11]. In some trials, the concentration of NaCl in the reaction mixture was varied ranging from 50 to 200 mM. At a low concentration of NaCl , e.g. 50 mM, the stainings were completely covered with choromogen deposits. At a NaCl concentration of 150 mM, very clear localization of the signals was obtained. However, the staining was suddenly lost at the higher concentration of NaCl [12]. It is necessary to decide on the best concentration of NaCl for each individual trial.

Control experiments

In order to verify the specificity of the signal, we conducted various control expriments. When the probe was omitted from the reaction mixture, no nuclear staining was observed. Also, when a section was reacted with the ds oligo-DNA probe in the presence of an excess amount (100-fold) of unlabeled ds oligo-DNA probe, the nuclear staining was markedly decreased. In addition, when a section was digested with RNase before the reaction with the labeled ds oligo-DNA probe, the nuclear staining was not altered, indicating that the nuclear staining was not due to nuclear RNA.

Applications

Localization of CRE-binding proteins

We synthesized an oligo-DNA with a consensus base sequence of CRE (TGACGTCA) with three TTA repeats at the 5', end. Since the CRE sequence is palindromic, the oligo-DNA was allowed to self-anneal and form ds DNA with three TTA repeats at both ends. The CRE ds-oligo-DNA was irradiated with UV light to form haptenic T-T dimers. When the haptenized CRE ds-oligo-DNA reacted with frozen sections fixed with 4% PFA in PBS followed by enzyme immunohistochemical localization of the T-T dimers, nuclei of intestinal epithelial cells and brain cells were heavily stained [10].

Localization of ETF

Since abnormal expression of epidermal growth factor receptor (EGFR) is frequently associated with cancer development, the analysis of EGFR gene expression at the transcriptional level as well as the transcript level is helpful in understanding the abnormal nature of cancer growth. We attempted to localize EGFR transcriptional factors, EGFR specific transcription factor (ETF), in the frozen sections of A431 human epidermoid tumor transplated into nude mice by southwestern histochemistry. As probes for southwestern histochemistry, (+) and (-) sequences of the DNA seqment (91 base pairs (bp)) including ETF regulatory element were synthesized, allowed to be annealed and then tailed by terminal deoxynucleotidyl transferase with digoxigenin (Dig)-11-dUTP. The sites of Dig were visualized enzyme-immunohistochemically with horseradish peroxidase-labeled anti-Dig. The 91 bp probe effectively detected a single ETF band with a molecular mass of 120 kD on a southwestern blot of the crude nuclear fraction extracted from A431 tumor cells.

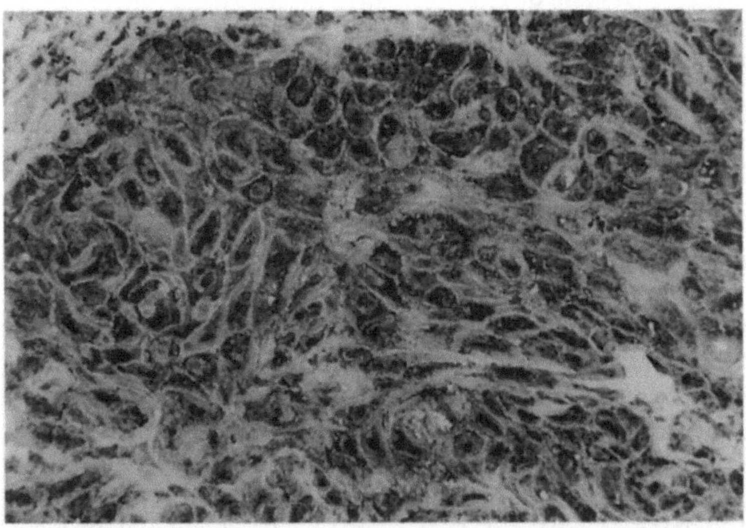

Fig. 1. Localization of ETF in frozen sections of A431 tumors by southwestern histochemistry. Using the 91 bp ds oligo-DNA probe, we attempted to localize the transcription factors of EGFR gene in the frozen sections of A431 tumors. When the sections were reacted with the 91 bp probe, the nuclei as well as the perinuclear regions were positive

When the frozen sections of A431 tumor were fixed with 4% paraformaldehyde and reacted with the 91 bp probe, the staining of perinuclear area as well as nuclei in a speckled pattern (Fig. 1) was observed and the staining intensity was increased depending upon the concentrations of the probe and reached a plateau level at 0.5-1 g/ml. Moreover, the nuclear staining with the probe was dependent upon a salt concentration and the signal/noise ratio was a maximum of 150 mM NaCl. The staining with the 91 bp probe was abolished by the presence of an excess amount of unlabeled 91 bp DNA or unlabeled ETF responsive element DNA alone, indicating that the nuclear and perinuclear staining with the 91 bp probe reflects the localization of ETF [12].

Conclusion

A histochemical method for the localization of specific DNA-binding proteins by using haptenized synthetic ds-oligo DNA with specific base sequences is introduced. Lists of RE and RE-binding proteins, including specific DNA-binding proteins that may be involved in the regulation of translation and/or mRNA stabilization, are rapidly growing. However, antibodies are not available for most of these newly identified binding proteins. Thus, southwestern histochemistry can be a novel tool to analyze cellular expression of gene-specific transcription regulatory factors.

References

1. Lin, C., R., Chen, W. S., Kruiger, W., Stolarsky, L. S., Wdber, W., Evans, R. M., Verma, I, M., Gill, G. N. and Rosenfeld, M. G.: Expression cloning of human EGF receptor complementary DNA: gene amplification and three related messenger RNA products in A431 cells. Science 224:843-848, 1984
2. Merlino, G. T., Xu, y, -h., Ishii, S., Clark, A. J. L., Semba, K., Toyoshima, K., Yamamoto, T. and Pastan, I.: Amplification and enhanced expression of the epidermql growth factor recptor gene in A431 human carcinoma cells. Science. 224:417-419, 1984.
3. Komuta, K., Koji, T., Izumi, S., Matsumoto, T., Kohara, N., Motojima, K., Kanematsu, T. and Nakane, P., K.: Expression of epidermal growth factor receptor messenger RNA in human colorectal carcinomas assessed by nonradioactive in situ hybridization. Eur. J. Surg. Onc., 21:269-275, 1995.
4. King, C., R., Kraus, M. H., Willoams, L, T., Merlinl, G. T., Pastan, I, H. and Anderson, S. A.: Human tumor cell lines with EGF receptor gene

amplificationin the absence of aberrant sized mRNAs. Nucleic Acids Res. 13:8477-8486, 1985.

5. Xu, Y. -H., Richert, N., Ito, S., Merlino, G. T. and Pastan, I.: Characterization of epidermal growth factor receptor gene expression in malignan and normal human cell lines. Proc.Natl.Acad.Sci.USA. 81:7308-7312, 1984
6. Nakane PK,Moriuchi T,Koji T,Tanno M,Abe K.: In situ localization of mRNA using thymine-thymine dimerized cDNA.Acta Histochem Cytochem. 20:229-43,1987.
7. Koji, T. and Brenner, R. M.: Localization of estrogen receptor messenger ribonucleic acid in rhesusmonkey uterus by nonradioactive in situ hybridization with digoxigenine-labeled oligodeoxynucleotides. Endocrinology 132:382-392,1993.
8. Silva, C. M., Tully , D. B., Petch, L. A., Jewell, C. M. and Cidlowski, J. A.: Application of protein-blotting procedure to the study of human glucocorticoid receptor interactions with DNA. Proc.Natl,Acad.Sci.USA. 84:1744-1749, 1987.
9. Koji, T., Izumi, S., Tanno, M., Moriuchi T. and Nakane, P., K.: Localization in situ of c-myc mRNA and c-myc protein in adult mouse testis.Histochem.J. 20:551-558, 1988.
10. Koji, T., Komuta, K., Nozawa, M., Yamada, S. and Nakane, P., K.: Localization of cyclic adenosine 3,5-monophosphate-responsive element (CRE) binding protein by In situ southwestern histochemistry. J.Histochem.Cytochem. 42:1399-1405,1994.
11. Slyden, O.D., Koji, T. and Brenner, R.M.: Microwave stabilization enhances imunocytochemical detection of estrogen receptor in frozen sections of macaque oviduct. Endocrinology. 136:4012- 4021,1995.
12. Komuta,K., Kanematsu T, Nakane, P., K and Koji, T.: Localization of Epidermal Growth Factor Receptor Enhancer Protein in A431 Epidermoid Carcinoma Cells by Southwestern Histochemistry. Acta Histochem Cytochem.31:267-273, 1998.

Subject Index

D

MIX
Papier aus verantwortungsvollen Quellen
Paper from responsible sources
FSC® C105338

If you have any concerns about our products,
you can contact us on
ProductSafety@springernature.com

In case Publisher is established outside the EU,
the EU authorized representative is:
Springer Nature Customer Service Center GmbH
Europaplatz 3, 69115 Heidelberg, Germany

Printed by Libri Plureos GmbH
in Hamburg, Germany